BOOST Your IMMUNE SYSTEM

Using God's Natural Remedies

Donald R. Hall, DrPH

Disclaimer: The ideas, concepts and opinions expressed in this work are intended to be used for educational purposes only. This work is sold and distributed with the understanding that the authors and publisher are not rendering medical advice of any kind, nor is this work intended to (a) replace competent and licensed professional medical advice from a physician to a patient, or (b) diagnose, prescribe or treat any disease, condition, illness or injury. Readers should seek the advice of their individual doctor or other healthcare provider prior to and during participation in these lifestyle activities. The author and publisher of this work disclaim any and all responsibility to any person or entity for any liability, loss or damage caused or alleged to be caused directly or indirectly as a result of the use application or interpretation of the material in this work.

Cover design by Gerald Lee Monks
Cover design resources from iStockphoto.com | Boonchuay1970 (Shoes) | ptasha (Excersize) | Mizina (Food)
Inside design by Aaron Troia

Additional copies of this book are available for purchase by calling toll-free 1-800-765-6955 or by visiting https://adventistbookcenter.com.

Library of Congress Cataloging-in-Publication Data

Names: Hall, Donald R., 1947- author.
Title: Boost your immune system : using God's natural remedies / Donald R. Hall, DrPH, CHES.
Description: Nampa, Idaho : Pacific Press Publishing Association, [2021] |
Summary: "A four-step strategy using natural remedies to boost the immune system and prevent infectious disease"— Provided by publisher.
Identifiers: LCCN 2020055012 (print) | LCCN 2020055013 (ebook) |
ISBN 9780816367306 (paperback) | ISBN 9780816367313 (ebook edition)
Subjects: LCSH: Natural immunity. | Communicable disease—Prevention. | Self-care, Health.
Classification: LCC QR185.2 .H35 2021 (print) | LCC QR185.2 (ebook) | DDC 616.07/9—dc23
LC record available at https://lccn.loc.gov/2020055012
LC ebook record available at https://lccn.loc.gov/2020055013

January 2021

Contents

Preventing Disease: A Four-Step Strategy

I've spent my life encouraging people to learn how to improve their health and prevent serious illness. It has always been challenging to motivate people to take a serious interest in improving their health. One positive outcome of the COVID-19 pandemic is that it focused people's attention on their health and how to stay well. The pandemic changed the whole world overnight. You couldn't visit friends, eat out, go to church, or in many cases, even go to work. Hopefully, it gets people thinking about what they can do to stay well.

To highlight the importance of good health, I often use an illustration to get people's attention. I take a million-dollar bill out of my billfold. (I got it at a dollar store.) Then I ask, "Who would like to have this million-dollar bill? I'll give it to you if you agree to be sick and miserable for the rest of your life." No one raises a hand. What good is a million dollars if you don't have good health? Then I ask, "How valuable is your health?" People quickly see that good health is their most valuable asset.

Too often, it takes a tragedy to get people to act on their health. While working in a hospital, I got a call from the president of a large corporation. He had heard about a wellness program I had started for hospital employees. The executive's message was simple, "I need your help! My top vice president just dropped dead from a heart attack, and I think I am next." It took the sudden death of his right-hand man for this corporate president to see that he needed to make some changes for the sake of his own health. I met with the executive, and we designed a six-month intensive wellness program for him and his top staff. We started with a comprehensive health evaluation. We even invited their spouses to participate because involving the whole family is helpful when talking about changing diets and lifestyles.

It was encouraging to see the whole group engage seriously in wellness activities, especially the company's president. He was right: he probably would have been next to have a heart attack. The president was overweight, smoked, had blood cholesterol over 300 mg/dL,[1] lived a stressful life, and worked long hours. During the six-month wellness program, he quit smoking, started a good exercise program, lost more than twenty pounds, changed to a primarily plant-based diet, and dropped his cholesterol level by a hundred points. At the end of the six months, the company president said he felt like a new man, and he was committed to maintaining his healthy lifestyle. In a letter thanking me for my help, the executive evaluated the wellness-program results of his staff. He said, "My managers look better, feel better, and work better!"

I think his brief statement correctly articulates what we all desire: to look our best, to feel good, and to be productive in life. That summarizes the goal of this book. By taking control of our health and lifestyle, we can go a long way toward achieving our life goals. Adopting a healthier lifestyle doesn't mean we will never be sick again, but it can greatly improve our odds of living longer, healthier lives. Harvard University conducted a large population study (123,000 people followed for thirty-four years) on preventing our nation's most serious health problems.[2] Consider the following benefits of people who lived a healthy lifestyle:

- Cut the risk of cancer by 65 percent
- Dropped the risk of heart attacks and strokes by 82 percent
- Decreased their risk of death from any cause by 74 percent
- Lived significantly longer—men 12.5 years longer, and women 14 years longer

These are amazing differences. Who wouldn't want to do the same in their lives? We can, but it takes a plan and a wellness strategy. In this booklet, we will lay out a plan to follow to help us achieve our best health. First, let's take a look at what causes disease. Then we'll consider the steps we can take to prevent disease and enjoy our best health.

THE CAUSE OF DISEASE

When you get sick, there are usually several reasons. If it is an infectious disease, such as COVID-19 or the flu, people usually emphasize the virus itself. However, other factors are involved in determining whether you get sick or not. Here is a list of common causes of most diseases, disabilities, and deaths:

- Pathogens such as bacteria or viruses (e.g., measles, chickenpox, tuberculosis, influenza, COVID-19)
- Genetics, which is a component of most chronic diseases, such as diabetes and sickle cell anemia
- Toxins and chemicals such as gasoline and household cleaners if taken internally, lead, asbestos, or an overdose of medicine
- Unhealthy lifestyle factors such as smoking, obesity, drinking habits, poor diet, lack of exercise, or poor sleep habits
- Impaired immune function caused by certain medications, fatigue, poor nutrition, or lack of sleep
- Injuries and disabilities caused by falls or accidents
- Mental or emotional issues such as excessive stress, dealing with loss, worry, hostility, loneliness, or a negative outlook on life
- Lack of professional health care, including regular health exams for blood pressure, high cholesterol, and high blood glucose, and checking for early disease while it can be successfully treated

As you can quickly see, there are many causes of illness, injury, and death. Thus, it is important to look at the causes and identify specific strategies to prevent or treat disease.

DISEASE PREVENTION

As we consider disease as a whole, I would like to emphasize four key steps you can take as a strategy to prevent illness and chronic diseases:

1. Limit your exposure to infectious diseases.

2. Maintain a healthy immune system.
3. Develop a healthy lifestyle. Apply God's natural remedies to prevent or manage chronic health problems that are largely preventable through lifestyle changes.
4. Get regular medical care. This includes managing existing health problems, getting recommended preventive exams and immunizations, and getting early medical care should you develop disease symptoms. Early detection and treatment greatly improve outcomes.

---

1. Total cholesterol for healthy adults should be under 200 mg/dL.

2. Yanping Li et al., "Impact of Healthy Lifestyle Factors on Life Expectancies in the US Population," *Circulation* 138, no. 4 (July 24, 2018): 345–355.

Step 1

Limit Your Exposure to Infectious Diseases

Infectious diseases are the most common cause of illness that most of us face regularly. They are caused by germs—organisms such as bacteria, viruses, fungi, or parasites. These organisms get into the body, multiply, and cause illness. Examples include colds, influenza, pneumonia, coronavirus, chickenpox, measles, HIV, sexually transmitted diseases (STDs), hepatitis, Lyme disease, and the list goes on.

In past years, infectious diseases were more common. In 1900, the top three causes of death in the United States were pneumonia and influenza, tuberculosis, and gastrointestinal infections. Diphtheria was the tenth leading cause of death. Fortunately, most of those diseases have been largely eliminated by vaccinations, sanitation, and other public health measures.

Many of these infectious agents are still around, however, and from time to time, new ones pop up and cause serious illness and death. For example, in the first eight months after COVID-19 came on the scene, more than 200,000 people died from this new virus. Each year the flu also comes around during flu season. It caused 12,000 to 61,000 deaths annually from 2010 to 2020.[1] Who knows what new disease may pop up next year?

Taking proactive steps to prevent these common causes of illness is important.

- Limit your exposure to these illnesses. Avoid people who are sick and contagious. From past experience, we've learned to wash our hands, sanitize surfaces, wear face masks, stay at a distance from people who may be ill, and stay home if sick to avoid spreading germs.
- Be sure you and your family members are vaccinated for

common illnesses as your doctor prescribes. Vaccines have virtually wiped out major killers such as smallpox and polio. Vaccines are still very important for childhood diseases, pneumonia, influenza, coronavirus, human papillomavirus (HPV), and other dangerous infectious agents, depending upon where you live.
- Practice good sanitation and cleanliness in your home and community.
- Limit contact with mosquitoes and ticks in areas where malaria, Lyme disease, and other diseases are passed from insect bites—stay out of the woods in tick-infested areas or wear insect repellent, use mosquito netting at night in countries where malaria is present, and take other recommended measures.
- Be sure your water supply is pure. If it's not, boil or purify it. The same applies to foods. In areas where intestinal infections are common, be sure food is thoroughly cooked, served on clean plates, and prepared by well-washed hands. Being proactive prevents most problems.

At one time, I worked for a month in Central America, where the World Health Organization (WHO) was working to lower infant mortality rates. In the village I was in, four out of every ten children died before reaching the age of five, primarily from intestinal infections. Sanitation was the key need in those homes. Washing one's hands helped the most.

MINIMIZING RISK

The first line of defense against illness is to limit one's exposure to germs. So what is the best way to do this? The following recommendations for minimizing the risk of getting or spreading infectious diseases were gleaned from the US government's Centers for Disease Control and Prevention (CDC) website at https://www.cdc.gov:

Wash your hands often

- Wash your hands often with soap and water for at least twenty seconds. It is especially important to wash before eating or preparing food, before touching your face, after using the restroom, after leaving a public place, after blowing your nose, coughing, or sneezing, after handling your face mask, after changing a diaper, after caring for someone who is sick, and after touching animals or pets.

- If soap and water are not readily available, use a hand sanitizer that contains at least *60 percent alcohol*. Cover all surfaces of your hands, and rub them together until they feel dry.
- Avoid touching your eyes, nose, or mouth with unwashed hands.

Avoid close contact with people who may be sick

- Inside your home, avoid close contact with people who are sick.
- If living with someone who has an infectious disease, wear a special mask (such as a plastic surgical mask) and gloves when caring for him or her. These masks are designed to filter out viruses when working in close contact with patients. Wash your hands thoroughly after contact.
- If you are sick, stay home

Cover coughs and sneezes

- Always cover your mouth and nose with a tissue or use the inside of your elbow when you cough or sneeze, and do not spit.
- Throw used tissues in the trash.
- Immediately wash your hands with soap and water for at least twenty seconds. If soap and water are not readily available, clean your hands with a hand sanitizer that contains at least 60 percent alcohol.

Clean and disinfect

- Clean and disinfect frequently touched surfaces. This includes tables, doorknobs, light switches, countertops, handles, desks,

phones, keyboards, toilets, faucets, and sinks. Some viruses can live on hard surfaces for up to four days.
- If surfaces are dirty, clean them. Use detergent or soap and water before disinfection. Then, use a household disinfectant. Most common Environmental Protection Agency (EPA) -registered household disinfectants will work.

Monitor your health

- Be alert for symptoms. Watch for fever, cough, shortness of breath, diarrhea, sore throat, or other symptoms of an infection.
- If symptoms develop, take your temperature; rest; eat light, healthy meals, and drink plenty of fluids; but if symptoms get significantly worse or if you are concerned, contact a doctor or a medical professional and follow his or her directions. If the symptoms listed above become severe, call 911 and get emergency help right away.

Doctors can test to see whether you have the flu or other infectious diseases. Most counties and states also have free testing available for anyone with symptoms of some infectious diseases. Check the internet for free testing and testing sites in your county.

The Centers for Disease Control and Prevention has an interactive screening tool for persons who think they may have COVID-19. You can find it on the web at https://covid19.apple.com/screening. If you have COVID-19, follow your doctor's instructions. You may also want to read the CDC guidelines "What to Do If You Are Sick," which can be found on the CDC website at https://www.cdc.gov/coronavirus/2019-ncov/if-you-are-sick/steps-when-sick.html.

---

1. "Influenza (Flu): Figure 1: Estimated Range of Annual Burden of Flu in the U.S. From 2010–2020," Centers for Disease Control and Prevention, last reviewed October 5, 2020, https://www.cdc.gov/flu/about/burden/index.html.

Step 2

Maintain a Healthy Immune System

The second defense against disease is to keep the immune system working efficiently. Many things decrease immunity, while others improve immune function. Many people catch an infection simply because their immune systems are impaired.

HOW THE IMMUNE SYSTEM WORKS

God designed the immune system to fight disease in the body, especially infections. After germs (viruses and bacteria) enter the body, they multiply and attack the body's cells. If the number of germs is low, the body can quickly overcome the infection in most cases. The larger the number of germs, the greater the risk that they may get the upper hand and cause illness.

Several immune system cells and tissues fight infections. In the blood, special cells called white blood cells (WBC) continually float throughout the body, looking for germs or foreign invaders.

There are several types of WBCs. Macrophages are WBCs that circulate in the blood and body tissues. When they recognize germs, they engulf and kill them. Then they release parts of the invading germs, called antigens, which alert other immune functions to begin.

B-lymphocytes or B-cells, another type of WBC, detect the antigens and go through a change to become plasma cells that begin making antibodies. When fully activated, each plasma cell can produce up to two thousand antibodies a second. Antibodies recognize the antigens on the germs and viruses and attach themselves, enabling the invaders to be destroyed by macrophages and neutrophils (another kind of WBC). The antibodies can also directly inactivate the germ and can prevent the virus from entering cells in the body. The first time the body is exposed to a germ, the body may take several days to gear up to fight the infection and produce maximum amounts of antibodies.

A third WBC type that destroys viruses is T-lymphocytes, or T-cells. T-lymphocytes identify body cells that have been invaded by viruses and have been turned into virus-making factories. When infected cells are found, the T-cells destroy them; T-lymphocytes also play an important role in finding cancer cells and destroying them.

Following an infection, the body produces T-cells and B-cells that have a "memory." If the body is infected with the same virus again, the immune system can quickly ramp up to destroy it before it gains a foothold and causes sickness.

Vaccines help develop immunity by imitating an infection without causing the disease. A vaccine initiates T-cells to recognize a specific virus. When the virus shows up again, the body can quickly mobilize the immune system to take care of the infection and prevent the person from becoming sick.

Another defense is the network of lymph nodes placed throughout the body. Lymph is a watery substance that leaks from blood vessels and coats all the cells of the body. The lymph is transported in small lymph vessels back to the blood supply. On the way, the lymph is filtered through lymph nodes, rich in lymphocytes, that catch germs and cancer cells and destroy them. When there is an infection, the lymph nodes swell from producing more WBC to fight the infection. Thus, they play an important role in our immune system. And there are two important ways to stimulate lymph flow and function, improving lymphatic system operation—drinking plenty of fluids, and staying physically active.

As you can see, a healthy immune system is essential for fighting infectious diseases. This is especially true for viruses. Activating your own immune system is the primary way the body recovers from virus infection. Loy Anderson, an Adventist physician, wrote a piece in *Burman University Magazine* in which she made this key statement about the immune system: "I know for sure that the only true defense is a strong and healthy immune system."[1] We can thank God that He gave us a marvelous immune system to fight against these dangerous viruses. We can also thank Him that He instructed us how to maintain good health and an effective immune system, as we will see in our study of God's natural remedies in the following pages.

FACTORS AFFECTING THE BODY'S IMMUNE SYSTEM

Clearly, an important strategy to consider in preventing disease is to have a strong, healthy immune system. Two people may be exposed to a virus. One gets sick, and the other doesn't. Why? It may be the sick person's immune system is not up to par, or he or she may already be weakened by an infection from another virus (cold, influenza, mononucleosis, HIV, etc.). Here are some factors that decrease the effectiveness of the immune system in fighting viruses:

- aging
- pregnancy
- lack of sleep
- poor nutrition—especially diets low in fresh fruits and vegetables or low in total calories and protein
- obesity and high blood sugar
- chronic inflammation
- low exercise level
- high stress levels
- lack of immunizations

Let's take a quick look at each of these important factors affecting immunity. More detailed information on these factors will be discussed in the section dealing with God's remedies.

Aging

As a person ages, the immune system simply isn't as effective as a younger person's because the number and quality of immune cells decrease. For example, this decreased immune response is obvious with the coronavirus. One study showed that in people over sixty years old, the COVID-19 virus is about a hundred times deadlier than in those under forty.[2] The reasons why are not fully understood, but one important factor is that levels of general inflammation are generally higher in people as they age.

In a study reported in the *Journal of the American Medical Association* (*JAMA*), researchers tracked the case-fatality rate of 1,625 patients in Italy who had COVID-19 disease by age.[3] Here is what they found.

Age	Case-fatality rate (%)
80+	20.2
70–79	12.8
60–69	3.5
50–59	1.0
40–49	0.4
30–39	0.3
Under 30	0

For now, the important takeaway is to keep older people more isolated from exposure to other sick people. This is even more important for those with existing health conditions, such as high blood pressure, obesity, lung disease, or diabetes. People with these ailments need to take positive steps to enhance their overall health and immunity.

Pregnancy

The Centers for Disease Control and Prevention website suggests that pregnant women have a greater risk of becoming severely ill with coronavirus, and they should be very careful.[4] They also can take action to enhance their immune system.

Lack of sleep

Lack of sleep can lower your immune system's efficiency and make you more susceptible to an infection from a virus. Numerous studies show that people who lack adequate sleep or experience poor quality of sleep are more likely to get sick when exposed to a cold virus. For the best immunity and overall health, adults should aim for at least seven to eight hours of sleep daily. A consensus statement by the American Academy of Sleep Medicine and Sleep Research Society states:

- Adequate sleep is essential for "optimal health."
- Sleeping less than seven hours per night regularly "is associated with adverse health outcomes, including weight gain and obesity, diabetes, hypertension, heart disease and stroke, depression, and increased risk of death."
- Sleeping less than seven hours per night "is also associated with impaired immune function."[5]

From personal experience, I've noticed over the years that when I catch a cold or the flu, it is often soon after I have stayed up late for a night or two. I've learned that burning the midnight oil is just not worth it. Regularity in sleep hours is a good health habit, especially for good immune function.

Poor nutrition

A diet lacking one or more nutrients can impair the production and activity of immune cells and antibodies. (See the section on healthy eating and foods that improve immune function and lower inflammation—"Proper Diet" in Step 3 of this booklet.) This is a big factor in preventing disease and combating any infection successfully.

Obesity and high blood sugar

Obesity and high blood sugar levels seem to go together. They are also common factors associated with an increased risk of contracting a disease. When blood sugar levels are high, the risk for all infections increases, as does generalized inflammation. The same is true of obesity. Again, there is much one can do to lower blood sugar levels and lose excess weight, both of which, in turn, lower the risk of disease. For example, Barry Popkin of the University of North Carolina at Chapel Hill is quoted as saying that if you contract the novel coronavirus, "You have more than double the likelihood of going into the hospital if you're obese and 50% more likelihood of dying."[6]

Chronic inflammation

Inflammation is an underlying factor in many major diseases, including diabetes, cancer, coronary heart disease, and high blood pressure. Excess inflammation makes it harder for the immune system to do its job of killing invading viruses. Doctors have discovered that giving a steroid drug—a potent anti-inflammatory—has significantly improved survivability in seriously ill COVID-19 patients, cutting death rates by 20 percent for patients on oxygen and by 35 percent for patients on ventilators.[7]

We also know that inflammation is more common in older people and people who are obese, who are inactive, who smoke, who eat diets high in animal foods, and who have high blood sugar levels (diabetes). All of these people are also at high risk for mortality from infectious disease.

Low exercise level

Physical activity has many benefits for protecting against disease. First, it improves immune function. How it does this is not fully understood, but many studies show that people who exercise regularly are less susceptible to colds and the flu. Regular exercise helps lower blood sugar levels, systemic inflammation, blood pressure, and the risk of obesity. Exercise is also important for lung health. All of these factors lower the risk of disease.

High stress levels

High levels of stress over time can adversely affect the immune system. The Cleveland Clinic published an article on its website about stress and the immune system; it states that when a person is under prolonged stress, the body produces high levels of cortisol, which makes the person vulnerable to higher inflammation, increased infections, depression, anxiety, and a depressed immune system.[8]

Lack of immunizations

Immunizations, when available, are one of the most effective ways to protect oneself from serious infectious diseases. Vaccines stimulate the body's immune system to recognize specific viruses and bacteria and thus protect the person from getting sick. When a person gets an infection and survives, his or her immune system develops a natural immunity. Being vaccinated accomplishes the same thing but without having the effects of the disease. Vaccines provide the best option for a nation to control serious infectious diseases such as the flu. Scientists estimate that at least 70 percent of the population must develop immunity (by either surviving the disease or being vaccinated) to achieve herd immunity and normalization.[9]

---

1. Loy Anderson, in "Engulfed! A World in Crisis," *Burman University Magazine* 3, no. 1 (Fall 2020): 5.

2. Nir Menachemi, "Coronavirus Is Hundreds of Times More Deadly for People Over 60 Than People Under 40," Conversation, September 10, 2020, https://theconversation.com/coronavirus-is-hundreds-of-times-more-deadly-for-people-over-60-than-people-under-40-145510.

3. Graziano Onder, Giovanni Rezza, and Silvio Brusaferro, "Case-Fatality Rate and Characteristics of Patients Dying in Relation to COVID-19 in Italy," *JAMA* 323, no. 18 (May 12, 2020): 1775, 1776, http://doi.org/10.1001/jama.2020.4683.

4. "Coronavirus Disease 2019 (COVID-19): Pregnancy, Breastfeeding, and Caring for Newborns," Centers for Disease Control and Prevention, updated November 3, 2020, https://www.cdc.gov/coronavirus/2019-ncov/need-extra-precautions/pregnancy-breast feeding.html.

5. Consensus Conference Panel of the American Academy of Sleep Medicine and Sleep Research Society, "Recommended Amount of Sleep for a Healthy Adult: A Joint Consensus Statement of the American Academy of Sleep Medicine and Sleep Research Society," *Sleep* 38, no. 6 (June 2015): 843, 844, http://doi.org/10.5665/sleep.4716.

6. Barry Popkin, quoted in Jennifer Abbasi, "Large Meta-analysis Digs Into Obesity's COVID-19 Risks," *JAMA* 324, no. 17 (November 3, 2020): 1709, https://jamanetwork.com/journals/jama/fullarticle/2772071.

7. Adam Feuerstein, "Inexpensive Steroids Reduce Deaths of Hospitalized Covid-19 Patients, WHO Analysis Confirms," *STAT*, September 2, 2020, https://www.statnews.com/2020/09/02/covid19-steroids-reduce-deaths-of-hospitalized-patients-who-analysis-confirms/.

8. "What Happens When Your Immune Systems Gets Stressed Out?" Cleveland Clinic, March 1, 2017, https://health.clevelandclinic.org/what-happens-when-your-immune-system-gets-stressed-out/.

9. Cynthia DeMarco, "COVID-19 Herd Immunity: 7 Questions, Answered," MD Anderson Cancer Center, last updated December 7, 2020, https://www.mdanderson.org/cancerwise/what-is-covid-19-coronavirus-herd-immunity-when-will-we-achieve-herd-immunity.h00-159383523.html.

Step 3

Develop a Healthy Lifestyle

The third step in our strategy for preventing both infectious and chronic disease is to develop a healthy lifestyle. People who keep fit with exercise; maintain a healthy weight; choose healthy foods; avoid tobacco, alcohol, and illicit drugs; get adequate rest; and maintain a positive, hopeful attitude have stronger immune systems. Additionally, they are much more likely to survive a serious disease should they have one. Existing health conditions (obesity, diabetes, heart disease, high blood pressure, lung disease, etc.) make recovering from any infectious disease much more difficult.

For example, data on the Our World in Data website, which provides "research and data to make progress against the world's largest problems," compared mortality rates in people who contracted COVID-19 with or without various health problems.[1] Here's what the research found:

Health problem	Mortality rate (%)
Cardiovascular disease	10.5
Diabetes	7.3
Chronic lung disease	6.3
High blood pressure	6.0
Cancer present	5.6
No other health problems	0.9

As you can see, if you have existing health problems, your risk of death is significantly increased when you contract an infectious disease. If you have multiple health conditions, your risk is even greater.

The COVID-19 pandemic taught us the importance of maintaining good health if we want to survive an infection. In a large study of the disease in New York (4,103 patients with the disease), researchers found that, next to age, obesity was the most common factor among those admitted to the hospital. "The chronic condition with the strongest association with critical illness was obesity, with a substantially higher odds ratio than any cardiovascular or pulmonary disease."[2]

Another early study in *JAMA* looked at case-fatality rates and the risk of hospitalized patients dying from COVID-19 when they had other existing health problems. Here is what they found:[3]

Health problem	Mortality rate (%)
Diabetes	36
Ischemic heart disease	30
Active cancer	20
Atrial fibrillation	25
No coexisting disease	0.8

From this and prior data, you can see that your risk of having a serious problem with COVID-19 is very low if you are under fifty and have no existing diseases. Fortunately, in the US, there is much you can do to prevent these common health problems, as you will see.

The personal experience of a young pastor shows how much a person can improve one's health. The pastor was significantly overweight and not particularly interested in following a healthy lifestyle. Then he caught COVID-19. He was seriously ill, was hospitalized, and recovered slowly over several weeks. The experience made him think about his lifestyle. He was not exercising, and he was very overweight and eating poorly. He began talking to his wife, who was a nurse, about what he could do to improve his health. Together they started him on a healthy weight-loss program. He began walking regularly and eating better, and over many weeks, he lost more than forty pounds.

That was when he contacted me to learn about resources to teach healthy living classes in his church and community. He expressed to me how much better he felt! He was excited about his new life. He signed up for a self-study program on health ministry I had prepared for pastors, and he became even more interested in following God's plan of healthy living. Now he is enrolled at the Loma Linda University School of Public Health, preparing for ministry to the whole person—helping people experience physical, mental, and spiritual health. His wife told me, "Thank God for COVID. It changed my husband's life!"

When God created man and woman, He made a special garden for them to live in. (You can read the account in Genesis 2.) The air was free from pollution. The water flowing through the Garden was pure and fresh. Their food was perfectly designed for them (Genesis 1:29). They were given work to keep them physically active (Genesis 2:15). In the evening, God would come and walk with them, and they got to know Him better. All was good. They had access to the tree of life, which perpetuated perfect health and immortality. Then they chose to disobey God and decide for themselves how to live—to learn about good and evil firsthand. Soon sickness and death became a part of their lives and thus a part of our daily lives today.

In the early days of Seventh-day Adventism, our church pioneers and people throughout the nation were often sick and unable to work. Something needed to be done. In a vision on June 6, 1863, just sixteen days after the church formally organized, Ellen White received a message from God on the importance of healthy living.

One of her early statements on health made this simple recommendation on following God's plan for a healthier life: "Pure air, sunlight, abstemiousness, rest, exercise, proper diet, the use of water, trust in divine power—these are the true remedies. Every person should have a knowledge of nature's remedial agencies and how to apply them."[4]

These were the same basic ingredients for good health that God had provided in the Garden of Eden. As time passed, God provided additional information on healthy living. This health message has been a great blessing to all who follow it.

Today, the Adventist Health Study has documented that Adventists, as a group, have the best longevity of any group of people that has ever been studied, and the study results were presented in a scientific health journal.[5]

The *National Geographic* magazine initiated a study to find where the healthiest and oldest people lived. They found five areas in the world, termed "Blue Zones," where people commonly lived over one hundred years and grew old without an abundance of health problems, such as heart disease, diabetes, and cancer. They described these zones as "where the world's healthiest people live." The community of Loma Linda, California, was designated as one of these Blue Zones. (It is the only one in North America.)[6] This community is largely Adventist and is where Loma Linda University (LLU) is located. LLU is a medical university that trains doctors, nurses, dentists, allied health professionals, and public health workers. The university's motto is "To Make Man Whole." Its faculty and staff teach healthy living, and they put the health principles into practice in their own lives.

I will summarize these key health principles that have been taught for more than a hundred years in the Adventist Church. They are practical suggestions on how you might improve your health today. Taken as a whole, they can significantly lower the risk of all diseases and, at the same time, help you live a longer, healthier life. Let us take a closer look at each one of these natural remedies.

PURE AIR

Scientists have found that to reduce the risk of spreading disease, it is far better to be outdoors than indoors because virus droplets disperse in fresh air. Viruses spread in the air in small water droplets expelled by a person's cough, sneeze, or even by talking. Wearing masks helps reduce the spread of these droplets, but being outside is even more protective because they quickly disappear in fresh air. Indoors, the virus droplets can accumulate, increasing the risk of infection.

Doctors recommend introducing fresh air to reduce the risk of infection. Open doors and windows. Let fresh air in. Doctors recommend a ventilation system in offices and classrooms that will change the air in the room several times an hour to reduce the risk of infection.

In a coronavirus health report from the United Kingdom, a member of the Scientific Advisory Group for Emergencies stated, "The science suggests that being outside in sunlight, with good ventilation, are both highly protective against transmission of the virus."[7] The report suggested walking in a park or having a picnic outside. Experts at the London School of Hygiene and Tropical Medicine state that particles of moisture that are holding viruses are diluted by fresh air, and they added that UV light from the sun also helps to destroy viruses.[8]

An infectious disease expert at the University of Nottingham said, "The use of outside spaces in a socially distant way is one of the lowest-risk forms of activity." He said he intended to go fishing and play golf.[9] Go for walks. Ride your bike. Enjoy the freedom of being outside and keep yourself healthy.

Researchers from the University of California, Davis, and the University of Oregon published a report in the journal *mSystems* (the journal of the American Society for Microbiology), making the following recommendations for healthier workspaces:

- Open windows for better air circulation.
- Open drapes and blinds to let in more natural sunlight.[10]

This applies to classrooms, offices, day care centers, assisted living facilities, universities, state offices, churches, and our own homes. These simple suggestions can have a significant impact on one's health. Take advantage of these two natural remedies, free from God, to enhance your health.

Benefits of fresh air

There are many more benefits of fresh air. Air pollution was estimated to have caused some 107,000 deaths in the US in 2011.[11] The Royal College of Physicians estimates that there are 40,000 deaths in the UK each year due to air pollution.[12] A UK physician group stated the following benefits of fresh air:

- *Increases energy*, especially when outside in a natural setting. Being out in the fresh air is a better energy boost than the traditional cup of coffee or energy drink. Fresh air and oxygen vitalize the blood, carrying more oxygen to the whole body. Fresh air is especially good for the brain, helping you stay sharp mentally.

- *Improves digestion.* After a meal, take a stroll in the fresh air to improve digestion.
- *Cleans lungs.* Everyone knows the importance of not smoking, the worst type of air pollution, resulting in more than 440,000 deaths annually in the US. Even in everyday air, there is some pollution: dust particles, viruses, or car exhaust. Deep breathing in pure air can help clear your lungs. Mucus traps dust particles and viruses. Deep breathing, and even coughing if needed, helps the lungs remove the pollution they have trapped. Drinking at least two quarts of water daily also helps make more mucus in your airways to remove impurities.
- *Relieves stress and anxiety.* When feeling anxious, stop and take a deep breath of pure air, preferably outside where you can smell the trees and flowers. The increased oxygen is thought to increase serotonin levels released in the body, in turn helping you feel happier and relaxed. Take several deep breaths, letting the air out slowly. Feel the tension going out of your body. It is a natural relaxer of the body and mind.
- *Bolsters the immune system.* Taking in more fresh air with exercise stimulates the immune system. It increases the natural killer cells in the blood (T-cells)—the special cells that round up and kill pathogens and viruses. The number of neutrophils and monocytes (white blood cells that kill germs) is also increased, boosting your immune system.[13]

With these benefits, it makes good sense to head outdoors often and breathe in nature's healing elixir—pure air. As the oxygen in your blood increases, every cell of your body will be benefited.

One obvious cause of impure air, as mentioned, is smoking. Smoking depresses the immune system and significantly increases the risk of disease. For example, one study in China found that those who smoked were three and a half times more likely to have a more severe experience with coronavirus disease than nonsmokers.[14] Smoking increases the risk for chronic diseases as well, including cancer, heart disease, stroke, chronic obstructive pulmonary disease (COPD), diabetes, and Alzheimer's disease. It is clear that you should avoid all tobacco if you want to decrease your risk for disease.

SUNLIGHT

Sunlight is essential for life. It is also a natural way to destroy viruses that can cause infection. One study looked at the viability of viruses in both dark and light rooms. The virus half-life was reduced from 32 minutes in a dark room to 2.4 minutes in a brightly lit room.[15] Let the sun in. It helps sanitize any room.

On a recent plane trip to see our grandkids, my wife and I wanted to minimize risk, so she brought a portable ultraviolet light that can be carried easily in her purse. Radiating the area around our seats killed any viruses, making our immediate environment more secure. Sunlight does the same when it is available. Some research showed that sunlight could destroy viruses in the small vapor droplets breathed out by people before they dissipate.[16] Sunlight is nature's sanitizer.

Sunlight has many more health advantages. On the skin, it makes vitamin D, which is an essential vitamin for good health. While vitamin D can also be found in some foods or supplements, the sun is the primary source of vitamin D for most people.

There has been much research on vitamin D and its effect on the immune system. Much is still unknown, but here are some things that we do know:

- All immune cells have vitamin D receptor sites, which confirms that vitamin D is important for proper immune function.
- Vitamin D has been shown to help activate T-lymphocytes, which play an active part in killing cancer cells, viruses, and cells that produce viruses in the body.
- Vitamin D is thought to help modulate the immune system,

helping prevent an excessive response of the immune system, thus lowering inflammation.

One large study looked at the vitamin D levels of populations in European countries and the risk of getting or dying from COVID-19. Researchers took the average vitamin D levels of populations in twenty European countries and correlated that with the number of cases and deaths from the disease. They discovered a significant relationship between vitamin D levels and cases and especially the mortality caused by this infection.[17] The most vulnerable population group is seniors, including nursing home residents. They are also the ones who are most deficient in vitamin D and at the highest risk. This was an association study, so it doesn't prove that lack of vitamin D *caused* the excess cases and death from COVID-19, but it provides evidence suggesting that getting adequate vitamin D may help the body deal with infectious diseases.

Although much is yet to be learned, vitamin D research scientists in the US have found that people with low levels of the so-called sunshine vitamin could be at greater risk of death from coronavirus.[18] We do know, however, that vitamin D has many health benefits beyond combating infections. It is especially important for the health of both bones and muscles, so make sure you get a moderate amount of sunshine regularly (avoid sunburns). If your vitamin D levels are still low, taking a supplement of vitamin D may provide health advantages. Typical supplements are about 2,000 IU daily, but daily intakes of up to 4,000 IU are considered safe. Ask your doctor for personal guidance.

As we all know, sunshine and pure air are wonderful physical and mental stimulants for everyone. It is hard to be sad when you are outside in a natural setting, feeling the warmth of the sun on your skin and enjoying pure, health-promoting fresh air.

ABSTEMIOUSNESS

The word *abstemiousness* is not used frequently today, so I searched online to get a better understanding. Here is a definition I found: "The quality of being abstemious, temperate, or sparing in the use of food and strong drinks. It expresses a greater degree of abstinence than temperance."[19] In other words, not eating or drinking too much. The Cambridge Dictionary further expands the meaning to not "drinking alcohol."[20]

This is a good principle for a nation where two-thirds (69 percent) of its adults are overweight or obese. In a study of more than four hundred thousand coronavirus patients, researchers found that obese people had a 46 percent greater risk of getting the disease than people who were not obese and a 113 percent greater risk of having to be hospitalized because of the disease. Obese people also had a 74 percent greater risk of ending up in intensive care and a 48 percent greater risk of dying from the virus.[21]

The study points out that the greater risk was due to two problems caused by obesity—an impaired immune system and a higher prevalence of obesity-related diseases, such as diabetes, coronary heart disease, and high blood pressure.

Maintaining a healthy weight by avoiding overeating is a particularly important strategy for surviving infectious diseases. Currently, 36 percent of Americans are obese, putting them at increased risk for disease whatever their age. WebMD reports the following additional health risks for those who are obese:

- Obesity increases the risk of blood clotting.
- Obese people have lower levels of adiponectin, a hormone that protects the lungs and helps keep blood vessels clean and open. With less adiponectin, obese people may be more likely to experience lung inflammation during infection, more blood clots resulting in heart attacks, strokes, and lung damage.
- People with obesity have more ACE2 receptors on the surface of cells in their bodies. ACE2 receptors are the doors viruses use to enter a cell and start producing more viruses.
- Obese people tend to be insulin resistant, which turns up the ACE2 receptors even more. Insulin resistance also is the underlying problem that results in diabetes.[23]

All of these factors increase the risk of contracting an infectious disease. So what can a person do? First, determine your body mass index (BMI), which shows whether you are overweight. For younger adults, a BMI greater than 25 indicates being overweight, and a BMI of 30 or higher indicates obesity. BMI is your weight relative to your height. Use the chart below to determine if you fall into one of these two unhealthy categories.

Height (inches)	**BMI 25+ (Overweight)**	**BMI 30+ (Obese)**
4'10"	119+	143+
4'11"	124	148
5'	128	153
5'1"	132	158
5'2"	136	164
5'3"	141	169
5'4"	145	174
5'5"	150	180
5'6"	155	186
5'7"	159	191
5'8"	164	197
5'9"	169	203
5'10"	174	209
5'11"	179	215
6'	184	221
6'1"	189	227
6'2"	186	225
6'3"	200	240

If you are currently overweight or obese, set a short-term personal goal of losing at least 5–10 percent of your body weight or about ten to fifteen pounds. Once you have reached that goal and maintained your weight for several weeks, then do it again if you still need to lose weight. Losing even ten pounds has significant health benefits.

Do not follow an extreme diet. Typical dieting seldom works. People go on a very restrictive diet for several weeks, lose weight, and then gain it all back in a few months. To manage weight long term, you need to change your eating habits. Some suggestions on what to change follow. Also, keep in mind that everyone should eat abstemiously—be careful not to overeat, even if you are not overweight. It is much easier to avoid excess weight gain than to lose excess weight.

The following suggested changes to your eating habits provide guidelines for weight management:

Keeping serving sizes small to moderate

The principle is to eat abstemiously, or sparingly. If needed, use a smaller plate. This helps most people to take less and eat less. Limit seconds. If you still feel hungry after a meal, just wait for five to ten minutes, then decide if you still need to eat more or not. It takes time for the body to recognize how much was eaten and for satiety to occur. Do not rush your meals.

Choose unrefined foods

A recent study looked at eating processed foods and weight gain.[24] A large group of people was randomized to eat either a processed- or unprocessed-food diet for two weeks. Subjects were instructed to eat as much or as little as they desired. The amount they ate was carefully recorded. Those eating the processed-food diet ate 508 more calories per day than those eating similar but unprocessed foods. They also gained an average of two pounds over the two weeks.

After two weeks, the groups switched diets. When those who had eaten the processed foods switched to unprocessed foods, they ate less and lost the two pounds they had gained. Remember, they were told to eat as much as they wanted.

When foods are processed, they are changed from their natural state. In most cases, they have added fats, sugar, salt, and other additives. Often, the vitamins, minerals, and fiber are also removed.

When eating processed foods, people simply eat more calories than they need. Cookies, pastries, ice cream, and other processed foods taste good, and we tend to overeat these foods. (When was the last time we overate steamed carrots or broccoli?)

One way of eating abstemiously is to choose unrefined, whole foods in moderate amounts rather than stuffing ourselves with processed and refined foods. Unfortunately, more than half of the typical American diet is highly processed food.

Examples of processed and unprocessed foods

Processed food	Unprocessed food
Sugar Frosted Flakes	Oatmeal, Shredded Wheat
White rice	Brown rice
Apple pie	Fresh apple
Hamburger, French fries, cola	Black bean patty, whole-wheat bun, vegetable juice
Chips or candy bar	Fresh veggies or a mix of nuts and dried fruit
Soft drink	Fresh orange or orange juice
Ice cream	Fresh fruit salad

Eat primarily plant foods

Plant foods include whole grains, fresh fruits, vegetables and salads, beans and lentils, and nuts and seeds. Plant foods, eaten primarily in their unrefined states, are the best foods for limiting calories and promoting good health. They are generally lower in calories and higher in fiber, resulting in good satiety. Eating plant foods is also good for the planet—a topic that we hear so much about today.

The Adventist Health Study at Loma Linda University studied the diet and lifestyle of nearly ninety-six thousand Adventists. They found that those who ate primarily plant foods (vegetarians and vegans) weighed an average of thirty pounds less than those of the same height who were not vegetarians.[25]

Choose a high-fiber diet

Research shows that if you eat high-fiber foods (twenty-five to more than thirty-five grams of fiber daily), you are more likely to maintain a healthy weight. This has been a major component of the Weight Watchers eating plan.[26] High-fiber foods tend to fill you up before you eat too many calories. They are also absorbed more slowly, which keeps your blood sugar levels from rising rapidly. High-fiber foods are low glycemic foods. By preventing sugar spikes in your blood from refined foods, fiber keeps your insulin levels lower. This prevents the rapid movement of the glucose in your blood into storage, which keeps you from getting hungry too soon (low blood sugar from high insulin) and helps prevent weight gain. Examples of high-fiber foods include beans and peas (the highest in fiber of any food group), oatmeal and other whole grains, fruit, leafy greens and other vegetables, and flaxseed and other nuts and seeds.

Eat meals at regular times

Avoid snacking on typical snacks between meals. The calories in snacks can add up quickly. Most snacks add 100–150 calories each, which, if eaten daily, can add up to ten pounds of extra fat each year if not burned off through exercise. If you get very hungry between meals, choose a healthy snack, such as fresh veggies, apple slices, half an orange, or a small handful of sunflower seeds or nuts.

Some people find that as they get older, they need fewer calories and that two meals a day work well for them, helping them keep their weight under control.

Add regular exercise to good eating habits

For weight management, exercise benefits in several ways. First, it burns extra calories you may have eaten that day. Second, exercise increases metabolism, which often drops when eating less. Third, it significantly reduces the health risks associated with being overweight. Exercise builds stronger muscles and bones and lowers your risk of heart disease, diabetes, and high blood pressure.

Everyone needs to gradually work up to at least half an hour of aerobic exercise daily, such as brisk walking, jogging, biking, swimming, or other active sports. For weight loss, the Institute of Medicine recommends people exercise for at least an hour daily.[27] You can break this up into two sessions of half an hour daily, or even three sessions of twenty minutes. Longer exercise periods not only burn more calories but specifically mobilize fat to burn in the activity.

Healthy tracking

One predictor of success in controlling one's weight is simply keeping daily written records. Set goals, then keep records of how you are progressing. Weigh yourself every day or at least weekly. Write your progress down in a journal or a calendar. Also, write down your miles walked or biked or your number of steps taken each day. Check to see how you are meeting your goals. Tracking progress keeps your mind focused on achieving your goals, and you are more likely to succeed.

Professional help

If you need further help, talk to your doctor, a health coach, or a nutritionist. It helps to have someone work with you and report on how you are doing.

Another aspect of abstemiousness is watching your alcohol intake. As mentioned above, the term expresses a greater degree of abstinence from alcohol. Alcohol adds calories (empty calories), which contribute to weight gain. It also impairs the immune system. Any amount of alcohol decreases one's immunity, but larger intakes have an even more significant effect on impaired immunity. Long-term and heavy drinking impair the immune cells in the lungs and can directly damage the cells that line the lung surface, making it more difficult to ward off infections.

For example, the World Health Organization (WHO) makes these statements about alcohol and preventing infection:

- "Alcohol has effects, both short-term and long-term, on *almost every single organ of your body.* Overall, the evidence suggests that there is *no 'safe limit'*—in fact, the risk of damage to your health increases with each drink of alcohol consumed.
- "Alcohol use, especially heavy use, *weakens the immune system* and thus reduces the ability to cope with infectious diseases. . . .
- "Alcohol alters your thoughts, judgement, decision-making and behaviour. . . .
- "Heavy use of alcohol *increases the risk of acute respiratory distress syndrome* (ARDS). . . .
- "Avoid alcohol altogether so that you do not undermine your own immune system and health and do not risk the health of others."[28]

The WHO article points out many other health concerns caused by alcohol consumption, such as increasing the risk for many types of cancer, liver disease, and depression. Additionally, drinking alcohol often plays a role in suicide, homicide, accidents, violence, abuse, and many other social issues leading to injury or death. Alcohol use is one of the top preventable causes of mortality in the world. The only preventable causes of death greater than alcohol use are poor diet, smoking, and lack of physical exercise. It is estimated that eighty-eight thousand to one hundred thousand people die prematurely from alcohol use each year in the US. Worldwide, WHO estimates that three million deaths every year result from alcohol use, and it is "a causal factor in more than 200 disease and injury conditions."[29] For the best health, choose healthy beverages that are free from alcohol.

REST

When it comes to your health, sleep and rest play important roles. Insufficient sleep adversely affects the immune system, leaving you susceptible to infectious disease. When your body is overly tired or exhausted, your immunity is impaired. I have personally noticed this. I am most likely to come down with a cold or the flu when I overdo and find myself short on sleep from staying up late. Regularly getting seven to eight hours of sleep is one of our best defenses against infection.

Here is what the National Sleep Foundation says about sleep and immunity: "To stay healthy, especially during the influenza season, get the recommended seven to eight hours of sleep a night. This will help keep your immune system in fighting shape, and also protect you from other health issues including heart disease, diabetes, and obesity."[30] The organization also points out that should you get an infection, being well rested and getting plenty of sleep daily is one of the best things you can do to help yourself recover more quickly.

If you have been short on sleep, the National Sleep Foundation recommends taking naps to help catch up on sleep. It recommends a twenty- to thirty-minute nap in the morning or afternoon, whenever you can fit it in, to boost immunity.[31]

A recent study by the University of Tübingen in Germany showed that getting adequate sleep has the potential to enhance the efficiency of the T-cell response.[32] Remember, T-cells play an important part in the body's early recognition of cancer and virus infections and ramping up the production of antibodies to fight the infection.

Rest is important for remedying physical tiredness but is just as important when emotionally exhausted or stressed. A *Psychology Today* blog post points out that stress makes us susceptible to illness and disease because stress severely depresses the immune system. The author states that some experts believe that "stress is responsible for as much as 90% of all illnesses and diseases."[33] That may be an overstatement, but stress certainly does play a big role in most diseases, including heart disease and infectious diseases.

I asked a friend of mine who is a gastroenterologist, "Of your patients who come to you with digestive complaints, in what percentage of them is the problem really caused by stress?" He thought a moment, then answered, "Probably about ninety percent of them. At least, stress and emotional turmoil made the condition worse." This illustrates the need that we all have for rest and relaxation to keep our lives in balance.

When under continual stress, the body produces hormones that severely depress the immune system. The special T-lymphocytes that fight infection are reduced in number, as are other WBCs that destroy viruses. Ultimately, the whole immune system is weakened, making the body more susceptible to contracting infection and illness.

I was talking to a researcher who worked with the immune system in animal research. The researchers randomly divided the test animals into a stress group and a control group (not stressed). Then they looked at changes that took place in the animals' bodies when stressed. The researcher told me that the way he determined whether those in the stress group were adequately stressed for the experiment was by looking at their lymph nodes. If they showed signs of atrophy, then he knew the animal had been adequately stressed. In other words, high stress causes the body's immune system—in this case, lymph nodes that filter out viruses and other germs—to actually show signs of deterioration and disfunction. Continual high stress levels do the same thing to humans, causing atrophy and deterioration of the immune system.

For several years, I taught a stress-reduction class in a hospital setting. The participants were primarily hospital staff. As you can imagine, hospital work can be very stressful. We talked about many strategies to reduce stress, but what they found most helpful was the progressive relaxation exercise we did together in the class each week.

Part of this rest principle is just taking time to relax throughout the day. You cannot be stressed and relaxed at the same time. Relaxation works like an antidote for stress. When feeling anxious, stop and take two or three deep breaths. Each time you do, sigh on the exhalation. Feel the muscles in your shoulders relax. Let life's problems slip away as you relax. Fill your mind with images of pleasant surroundings. Remember a walk in the woods or a time you were sitting by the ocean, watching the waves come and go. Feel at one with nature. As you replace worry and stressful thoughts with moments of relaxation and beauty, you begin to lower tension and reduce stress hormones in your body. This boosts your immune system a little.

There is a stress-modifying practice called mindfulness. It is learning to slow down a bit and enjoy the world around you rather than filling your mind with all the problems of the day. You have heard the phrase

"Take time to smell the roses." There are beautiful things all around us. Take time to notice and enjoy them: the blue sky, the warm sun, a bird singing in a tree, and trees and flowers. Take time for a relaxing walk outside in nature. It is healing. Give thanks for the many blessings you enjoy.

Jesus talks about this simple concept of not stressing about what might happen tomorrow but focusing on the good things of today. "So don't be anxious about tomorrow. God will take care of your tomorrow too. Live one day at a time" (Matthew 6:34, TLB).

God also built into each week a special day (Exodus 20:8–11) to put aside work—a time when you can just rest and enjoy that day. He calls this Sabbath. A time when you do not have to worry about any work but just spend time with Him and your family; it is a day to spend time in nature, contemplating His gift to us. He knew we could not be revving our motors all the time. We also need regular times to rest. Take a Sabbath day's rest each week. Enjoy the break. It refreshes you for the next week and helps keep you healthy.

Improving sleep

If you have trouble sleeping or relaxing, the Mayo Clinic makes these suggestions for improving sleep:

- *Stick to a sleep schedule.* Go to bed and get up at the same time each day. The body loves consistency. By sticking to your schedule, you develop a sleep habit that will help you sleep better every night. If you do not fall asleep within twenty minutes, get up. Go into another room and do something relaxing. Read a book. Drink some chamomile tea. Listen to relaxing music. When you get sleepy, return to bed.
- *Pay attention to what you eat.* Avoid big meals late in the day. You need to go to bed on a mostly empty stomach. Lying down with a lot of food in your stomach causes gastroesophageal reflux disease (GERD, also known as heartburn), causing you to toss and turn and not sleep well. Do not drink coffee or other caffeinated beverages late in the day either. The

caffeine can keep you up. Also, if you drink, avoid alcohol. It may make you sleepy but can disrupt sleep later in the night.

- *Create a restful environment.* Make your bedroom comfortable for sleeping. This usually means a cooler room that is dark and quiet. Do not use light-emitting screens just before retiring. If your bedroom is not as quiet as you like, try earplugs. As sleep time approaches, switch to more calming activities—not an exciting TV show. Read for a while. Listen to calming music. Relax in a tub of warm water, if needed. It is a great way to relax and prepare for sleep.
- *Do not nap late in the day.* A nap earlier in the day is fine, but even then, limit the nap to fifteen to thirty minutes. If you work at night, a nap in the evening before going to work is fine.
- *Be physically active every day.* Exercise helps relax the body and burns up stress hormones. People who exercise daily go to sleep faster and sleep longer than those who do not exercise. Solomon, the wise man, points out, "People who work hard sleep well, whether they eat little or much. But the rich seldom get a good night's sleep" (Ecclesiastes 5:12, NLT). When you exercise, just be careful not to exercise strenuously just before bedtime. It may take your body time to ease down and become sleepy right after vigorous exercise.
- *Resolve worries before going to bed.* Do not tackle big problems just before bedtime. Resolve problems and worries beforehand. If you have a big problem on your mind, jot down what concerns you, then set it aside for tomorrow. If you have problems doing this, claim the Bible promise, "Give all your worries and cares to God, for he cares about you" (1 Peter 5:7, NLT), and follow this guidance from the apostle Paul: "Don't worry about anything; instead, pray about everything. Tell God what you need, and thank him for all he has done. Then you will experience God's peace, which exceeds anything we can understand. His peace will

guard your hearts and minds as you live in Christ Jesus" (Philippians 4:6, 7, NLT).

- *Ask your doctor for help.* Everyone has trouble sleeping at times. But if it continues night after night, ask your doctor for help. There may be an underlying cause that needs to be addressed to help you get the sleep you need for the best health.[34]

EXERCISE

Being physically active daily is one of the best health promoters for your body and mind. It improves overall fitness and the health of all cells in the body, including the immune system cells. One study showed that women who walked briskly every day for at least thirty minutes had only half the number of colds during the year compared to those who did not walk regularly.[35] Researchers are not sure how exercise helps the immune system, but the results are clear—those who exercise experience fewer infections from viruses.

Research shows that moderate exercise is best for stimulating the immune system. If the exercise is strenuous and sustained, the immune system may weaken, making one more susceptible to infections. This has been observed in marathon runners who push themselves hard to compete in long races. Thus, for good general health, participate regularly in moderate activities, such as brisk walking, biking, easy jogging, swimming, and other active sports.

Notice Ellen White's statements over a hundred years ago regarding exercise and health: "There is no exercise that will prove as beneficial to every part of the body as walking. Active walking in the open air will do more for women [or men], to preserve them in health if they are well, than any other means."[36] "God designed that the living machinery should be in daily activity; for in this activity or motion is its preserving power."[37] "The more we exercise, the better will be the circulation of the blood."[38]

A recent consensus statement by health and fitness professionals emphasizes the importance of moderate activities. "There is a general consensus that regular bouts of short-lasting (i.e., up to 45 minutes) moderate-intensity exercise is beneficial for host immune defense, particularly in older adults and people with chronic diseases."[39] Notice in the statement that exercise is especially beneficial to "older adults" and those with "chronic diseases." If you have diabetes, obesity, a heart problem, or are a senior, a walking program will be especially important to lower your risk of infection and chronic disease. If you have a chronic health problem, be sure to clear your activity level with your physician and ask for specific guidance to keep your exercise safe and effective.

Exercise improves the immune system and reduces the risk of many chronic diseases, such as diabetes, heart disease, hypertension, and obesity. An article in *JAMA Internal Medicine* clearly illustrates how lifestyle risks can increase mortality rates. The study included nearly four thousand individuals who were admitted to a hospital with serious coronavirus disease. Notice how the following risk factors increased mortality compared to people who did not have the risk factor.[40]

Risk factor	Increased mortality from COVID-19
High blood pressure	68%
High cholesterol	90%
High blood sugar (diabetes)	66%

The good news is that exercise helps prevent and correct these common risk factors.

Exercise and diabetes

Harvard University studied seventy thousand women over ten years. All were free of diabetes at the start of the study. Women who exercised forty to sixty minutes most days of the week had only half the risk of developing diabetes after eight years, compared to women who didn't exercise regularly.[41] Exercise works much like insulin to lower blood sugar. If you already have diabetes, exercise is a good way to lower your blood A1C levels.

Exercise and obesity

A study in the UK found that people who walked to work regularly had a 20 percent reduced risk of obesity and a 40 percent decreased risk of having diabetes. Those who biked had a 37 percent decreased risk of obesity and a 50 percent lower risk of diabetes.[42] Physical activity is also one of the best ways to burn extra calories and help you maintain a healthy weight for a lifetime. Of course, for those who are overweight, it is also helpful in losing those excess pounds.

Exercise and high blood pressure

Moderate exercise, such as brisk walking, is one of the best ways to prevent high blood pressure and help reduce it if it is too high. As you walk, your arteries open up, allowing more blood flow to the muscles. This helps relax the arteries and lowers blood pressure. The American College of Sports Medicine, in its guidelines for exercise, states that exercise should be a cornerstone for the prevention and control of high blood pressure.[43] It also points out that moderate activities, such as walking and biking, are most effective in lowering blood pressure. It also points out that weight training for strength helps lower blood pressure levels too.

Exercise and high cholesterol

When exercising, muscles use fat in the blood for energy, along with stored glycogen. This burns up fat circulating in the blood. With an hour of exercise daily, such as walking, you can lower the amount of fat in the blood by 40 percent or more, even in a couple of days (compared to no exercise days). Exercise lowers atherogenic triglycerides and LDL (low-density lipoprotein) cholesterol and lowers your risk of heart disease. At the same time, exercise raises HDL (high-density lipoprotein) cholesterol levels. HDLs help clear excess cholesterol from artery walls and thus greatly reduce one's risk of coronary heart disease. Exercise also lowers intra-abdominal fat, which produces LDL cholesterol that tends to promote fatty plaque in the arteries (atherogenic). In other words, regular exercise changes the way the body handles fat, lowering atherogenic fat particles and raising protective

HDL cholesterol. Both are protective of the heart.

Exercise is one of the most effective ways of lowering the systemic inflammation that increases the risk of chronic diseases, such as high blood pressure, cancer, diabetes, and heart disease. Exercise contributes even more directly to disease prevention by promoting good circulation, which allows the WBCs in the immune system to move through the body freely and do their job efficiently.

As you can see, exercise is an especially important part of your strategy for improving immunity and decreasing the risk of chronic diseases, which are the major cause of early mortality. The most common chronic diseases and their rate of prevalence are

- hypertension (46 percent);[44]
- obesity (42 percent);[45]
- diabetes (13 percent), and another third of the population have prediabetes (elevated blood sugar levels putting them at increased risk);[46] and
- coronary heart disease (6.7 percent).[47]

PROPER DIET

Harvard's Nutrition Source website makes this general statement about the foods we eat and immunity: "A balanced diet consisting of a range of vitamins and minerals, combined with healthy lifestyle factors like adequate sleep and exercise and low stress, most effectively primes the body to fight infection and disease."[48] Immunity is influenced by many factors, and diet is a major one. But what is a balanced diet?

A balanced diet is one that includes all of the nutrients the body requires for good health (vitamins, minerals, phytonutrients, fiber, etc.) and avoids high levels of food items that have an adverse effect on health (processed and refined foods, sugars, saturated fats, salt, and too many calories).

Nutrition and the immune system

First, let us briefly review what factors depress the immune system. Here is a list from Harvard's Nutrition Source website:

- older age
- polluted air (including smoking) and water that contains toxins
- excess weight
- poor diet or malnutrition
- chronic diseases, including inflammation
- chronic mental stress
- lack of sleep and rest[49]

As you can see, poor diet is just one of many factors affecting the immune system. The important point here is that all of these aspects of good health need to work together for good immunity. You can have the best diet in the world, but if you are significantly overweight, inactive, and not getting adequate rest, your immune system will still be significantly depressed.

Eating all the required nutrients found in a varied, unprocessed diet is required for the health and function of all cells of the body, including the immune cells. Nutrients identified from research "as critical for the growth and function of immune cells include vitamin C, vitamin D, zinc, selenium, iron, and protein."[50] Eating a diet limited in variety or a diet that is composed primarily of ultraprocessed foods, such as the typical Western diet high in refined sugars and red meat and low in fruits and vegetables, is associated with suppressed immunity.[51]

Additionally, we are learning that maintaining the health of the gut is also essential for good immunity. Our gut is home to trillions of microorganisms, which are referred to as the microbiome. Scientists are finding out that the microbiome also plays an important role in producing key substances that promote good immunity. The foods we eat determine how healthy the microbiome is and how well it functions. "A high-fiber plant-rich diet with plenty of fruits, vegetables, whole grains, and legumes appear[s] to support the growth and maintenance of beneficial microbes. Certain helpful microbes break down fibers into short-chain fatty acids, which have been shown to stimulate immune cell activity."[52]

Compare that statement to some that Ellen White wrote over a hundred years ago about the proper diet: "Grains, fruits, nuts, and vegetables constitute the diet chosen for us by our Creator. These foods, prepared in as simple and natural a manner as possible, are the most healthful and nourishing."[53] "The grains, with fruits, nuts, and vegetables, contain all the nutritive properties necessary to make good blood."[54] These statements on proper diet, along with the text in Genesis 1:29, have formed the foundation of the Seventh-day Adventist Church's teaching on encouraging a plant-based, whole-foods diet for the best health.

Choosing a variety of plant-based, minimally processed foods, in the proper amounts to maintain a healthy weight, is essential for good immunity and good health. Certain groups of people are at higher risk of eating an inadequate diet and thus have a higher risk of poor immunity. Here are the groups suggested by Harvard's Nutrition Source:

- *Elderly persons.* Seniors may have a poor appetite due to loneliness, depression, chronic health problems, lower interest in cooking for one person, or lack of access to good food due to economic reasons or transportation problems. It is estimated that about one-third of the elderly in industrialized countries have nutrient deficiencies.
- *Persons with very poor eating habits.* Detrimental habits include eating largely from fast-food outlets—hot dogs, hamburgers, beef tacos, fried chicken, soft drinks, pastries, snack foods—combined with a low intake of fruits and vegetables or a large intake of alcoholic beverages.[55]

These people may be helped by taking vitamin and mineral supplements. However, a better approach would be to eat healthier, well-balanced meals. The "Healthy Eating Plate" concept, provided by the US government and expanded by Harvard's Medical School and School of Public Health, provides guidelines for a balanced, healthy diet that meets nutritional needs for good health and good immunity.

The Healthy Eating Plate concept recommends that half of every plate should be fruits and vegetables. These foods are low in calories and high in vitamins, minerals, and phytochemicals associated with better health. The other half should be made up of whole grains and healthy proteins. Here are some guidelines for better nutrition:

- *The more veggies and the greater the variety, the better.* Aim for at least three to five half-cup servings daily. Eating more servings shows even greater health advantages. French fries and ketchup do not count!
- *Eat plenty of fruits of all colors.* Aim for at least two to three servings daily—more is better.
- *Eat a variety of whole grains.* Choose whole-grain bread and cereals, brown rice, whole-grain pasta, and oatmeal. Limit refined grains, such as white bread and rolls, white rice, and refined, sugary breakfast cereals.
- *Choose healthy plant proteins.* Eat plenty of beans, lentils, garbanzos, tofu and other soy foods, nuts, nut butters, seeds, and vegetarian entrées. Limit red meat and cheese. Avoid bacon, cold cuts, and other processed meats. If you choose to eat meat, fish and skinless poultry are healthier choices than red meat. Aim to have healthy, protein-rich food at every meal.
- *Use healthy fats.* Choose vegetable oils, including olive and soy oils, for cooking and on salads and other healthy fats, such as avocados, nut butters, flax meal, and chia seeds. Aim for at least one healthy fat at every meal.
- *Drink water as your primary beverage.* Limit dairy milk to one to two eight-ounce servings per day and pure juice to one small glass per day. Avoid sugary drinks and alcohol.
- *Stay active every day.* Exercise helps burn excess calories to maintain a healthy weight.

If the diet is deficient in even a single nutrient, it can diminish the body's immune response. That is why a healthy, balanced diet is so essential. Research has shown that if any of the following nutrients are low in the diet, one's immunity can diminish:

Zinc

Zinc is necessary for over one hundred enzymes in the body to carry out their vital reactions. Zinc is a major player in the creation of DNA, growing cells, building proteins, healing damaged tissue, and supporting a healthy immune system. A recent study in Spain looked at blood zinc levels and the rate of survival for hospitalized coronavirus patients. The average zinc level in those who died was significantly lower (43 mcg/dL) than those who survived the illness (63 mcg/dL).[56]

Zinc is also involved in taste and smell. You can quickly see why adequate zinc is important for combating infections. Good sources of zinc include legumes (peas, beans, lentils, and garbanzos), nuts and seeds (especially squash or pumpkin seeds, hemp seeds, and flax seeds), whole grains (quinoa, whole wheat, brown rice, and oatmeal), and dark chocolate.

Selenium

Selenium is a trace mineral, meaning you need only small amounts. Selenium helps make DNA and protects against cell damage and infections. The best food sources of selenium are Brazil nuts (one nut meets the daily requirement), whole-wheat bread, oatmeal, beans and lentils, eggs, sunflower seeds, dark chocolate, and fish.

Iron

Iron is needed for the normal development of the immune system and an adequate immune response when an infection occurs. Iron is needed for immune cell proliferation and maturation, particularly lymphocytes, when an infection is present. Getting adequate amounts is important, but getting too much (primarily from iron supplements or red meat) can decrease immunity. Good sources of iron from plant sources include legumes (soybeans are the best plant source, lentils, and tofu), whole grains (amaranth, oats, and quinoa), nuts (cashews, pine nuts, and almonds), seeds (pumpkin and flax), and leafy greens. Your body will not accumulate too much iron when getting it from plant sources. Plant sources of iron are absorbed better when eaten at the same meal with foods high in vitamin C (oranges, grapefruit, peppers, tomatoes, and most fresh fruit).

Copper

Copper helps keep the immune system healthy. Good food sources include nuts (almonds and cashews), sesame seeds (one tablespoon gives you nearly half of your daily requirement), sunflower seeds, avocados, leafy greens (spinach, Swiss chard, and kale), and dark chocolate.

Folic acid

Folic acid is one of the B vitamins and is essential for health. Folic acid helps make and repair DNA. Thus, it is vital for the production and maintenance of our body's cells, including immune cells. Good dietary sources include legumes (lentils, peas, beans, and soy), vegetables (especially asparagus, leafy greens, beets, Brussels sprouts, and broccoli), fruits (especially citrus, papaya, bananas, and avocados), and wheat germ.

Vitamin A

Vitamin A is needed to stimulate the production and activity of WBCs, which are the primary immune cells that protect your body from virus infection. There are two forms of vitamin A. The first is preformed vitamin A or retinol, which is found only in animal foods, and the other is provitamin A, such as beta-carotene, which is converted into retinol in the body. Beta-carotene is found primarily in yellow and orange fruits and vegetables. The best sources are leafy greens (kale, spinach, and broccoli), bright-colored vegetables (carrots, sweet potatoes, red bell peppers, winter squash, and tomatoes), yellow fruits (mangoes, cantaloupes, and apricots), and eggs. Vitamin A is good for many other aspects of health, such as healthy skin (also important to prevent infections) and healthy eyes. If you take supplements, it is best to take beta-carotene. Too much preformed vitamin A (retinol) can have toxic effects and increase the risk for osteoporosis. There are no toxic effects, however, when eating plant sources of provitamin A (beta-carotene).

Vitamin B_6

Vitamin B_6 is needed to support immune function. This function was shown in an experiment with critically ill patients in intensive care. Patients were randomly assigned to one of two groups: one received vitamin B_6 daily, and the other a placebo. After two weeks, those getting a placebo saw no change in their immune system. But those receiving vitamin B_6 showed an increase in T-lymphocytes and T-helper cells—positive changes that improved the patients' ability to fight infection, such as viruses.[57]

Good food sources of vitamin B_6 include chickpeas, potatoes, carrots, banana, edamame, marinara sauce, spinach, sweet potatoes, green peas, avocados, tofu, pistachios, fish, and winter squash. About 24 percent of people in the US who do not take supplements have low vitamin B_6 levels. Other people with low levels are those with celiac disease, ulcerative colitis, and those who are alcohol dependent.[58]

Vitamin C

Vitamin C is important for controlling infection and healing wounds. It is also a powerful antioxidant that helps protect cells from dangerous free radicals. Vitamin C stimulates the activity of WBCs; thus, it is important for fighting infections. The lack of vitamin C results in scurvy—a disease common among sailors in years past before we knew about vitamin C. Early sailors learned that they could prevent scurvy by taking limes on their ships for long voyages. Sailors even picked up the name of "limeys."

Another interesting historical note has to do with the Pilgrims who came to America on the *Mayflower*. During the winter, they had no fresh foods with vitamin C, so many developed scurvy and, due to low immunity, caught infections. About half of them died that winter. If only they had known that the wild berries and rose hips available in the new land could have given them plenty of vitamin C and prevented this early American tragedy.

Good sources of vitamin C include fresh fruit, such as oranges, lemons, papaya, apples, peaches, melons, berries, and kiwis, and fresh vegetables, such as peppers, tomatoes, parsley, leafy greens, broccoli, potatoes, and Brussels sprouts. Vitamin C and overall good nutrition make your body more capable of preventing infection and, should you get sick, to overcome your illness sooner.

Vitamin D

Vitamin D is both a nutrient (coming from food) and a hormone made by the body with the help of sunshine. Vitamin D has many functions in the body but specifically helps control infections and reduce inflammation. To see whether vitamin D would lower respiratory infections, researchers gathered data on more than ten thousand people taking vitamin D in randomized controlled trials. Those taking vitamin D supplements saw a significant reduction in acute respiratory infections.

In subgroup analysis, those people who had low vitamin D levels at the start of the study (25-hydroxyvitamin D levels less than 25 nanomoles per liter [nmol/L]) showed the most benefit from taking vitamin D. They had a 70 percent lower rate of respiratory infections compared to those not taking vitamin D. Those people with vitamin D levels above 25 nmol/L at the start of the study still showed lower rates of upper respiratory infection—a 25 percent reduction—when they took vitamin D supplements.[59] This study gives good evidence that adequate vitamin D is helpful in maintaining a healthy immune system and preventing infections.

In another study, levels of vitamin D were monitored in 235 hospitalized coronavirus patients. Patients who had sufficient vitamin D in their bodies—a blood level of at least 30 nanograms per milliliter (ng/mL)—had a significantly lower risk for serious complications, including losing consciousness, low blood oxygen levels, and death. In persons older than forty, those whose blood levels of vitamin D were sufficient were 51.5 percent less likely to die compared to those who were vitamin D deficient.[60]

In an earlier study led by Dr. Michael Holick, researchers found that having a sufficient amount of the sunshine vitamin (vitamin D) reduced the risk of becoming infected with coronavirus by 54 percent. They also observed that in those with adequate vitamin D, there was an increase in lymphocytes—immune cells in the blood that destroy viruses. This is an amazing finding and clearly illustrates the value of maintaining a healthy immune system by keeping your vitamin D level in an optimal range—a blood level of at least 30 nanograms per milliliter.[61]

So what actions should a person take? First, talk to your doctor about getting a test for your vitamin D level. That is the only way to know for sure if you are low or not. A large part of the adult population is low in vitamin D. The National Health and Nutrition Examination Survey, including 4,495 people across the US, found that 41.6 percent of the US population was deficient in vitamin D. Vitamin D deficiency was defined as a serum 25-hydroxyvitamin D concentration less than or equal to 20 ng/mL (50 nmol/L).[62]

After learning about the importance of vitamin D for good health, I asked my doctor to test my blood. Sure enough, I was low, too, even though I walked and jogged outside for more than an hour each day. My doctor told me that when he checked his levels, he was low too. So it is good to get it checked. If it is low, your doctor will tell you how best to get it into the healthy range. I now take 2,000 IU of vitamin D daily to keep it in the healthy range. Being out in the sunshine daily also helps. Having adequate levels of vitamin D helps prevent all types of infection in the respiratory system, such as the flu virus.

Vitamin E

Vitamin E is a very powerful antioxidant and enhances the immune responses to confer protection against several infectious diseases.[63] The primary source of vitamin E is plant oils (e.g., soybean, sunflower, corn, and wheat germ oil). There are two forms of vitamin E: alpha-tocopherol (the primary form in vitamin E supplements), and gamma-tocopherol (the primary form found in food and especially in soy and corn oils). Both forms of vitamin E seem to have protective effects on the body, but new research suggests that the gamma-tocopherol found in food is better for the health than alpha-tocopherol found in most supplements.

Human studies show that increased dietary intake of vitamin E improves immunity by increasing the number of lymphocytes (WBCs that destroy viruses in the body), increasing immunoglobulin levels and antibody levels (to help fight viruses), and increasing natural killer cells' activity that can destroy cancer cells and viruses.[64]

Good sources of dietary vitamin E include almonds, hazelnuts, sunflower seeds, wheat germ, mangoes, avocados, peanuts, broccoli, spinach, butternut squash, bell peppers, olives, Swiss chard, tomatoes, Brussels sprouts, kiwi fruit, soy foods, and plant oils—especially soy, wheat germ, and corn oils.

Being deficient in any of the nutrients listed above can increase your risk of a viral infection. That is why it is so important to have an overall healthy diet that is high in unprocessed foods and composed primarily of plant foods.

Foods high in cholesterol—red meats, butter, cream, and cheese—increase general inflammation in the body, decrease the efficiency of the immune system, and increase the risk of many health problems, specifically high blood cholesterol levels, higher blood pressure, higher blood sugar levels resulting in diabetes, and coronary heart disease. Eating a primarily plant-based diet of whole, unrefined foods helps your immune system to function at peak performance and helps prevent many of our nation's most serious health problems.

THE USE OF WATER

Water is the next item in the list of natural remedies. Water is essential for life and good health, both inside and out. First, water is a cleanser. Washing your hands and clothing and cleaning surfaces are all examples of how water helps keep us clean and free of viruses and bacteria that can make us sick. Lack of cleanliness is a major cause of disease, especially in underdeveloped countries where it is difficult to stop the spread of germs and viruses. Use plenty of water, with soap, to keep your body and home clean. This goes a long way toward preventing disease.

Hygiene and cleanliness are especially important when traveling. It is easy to pick up a new bug while on a trip, especially to new countries. Hotel rooms, restaurants, planes, and buses may be infected. When you travel, you may want to take measures to prevent contamination from other travelers. My wife and I recently took a trip on an airplane. We not only wore face masks but also took cleaning wipes to clean the armrests and areas around where we were sitting. It is prudent to make every attempt to cleanse your area, as it may have been contaminated by previous travelers.

Just as water is important for cleaning the outside of the body, water also helps keep the inside clean and functioning optimally. The liver and kidneys break down toxins, remove wastes, and purify the blood. All the blood passes through the kidneys many times a day. The blood transports oxygen to the cells and carries away carbon dioxide and other wastes from the cells. The blood also contains many WBCs (monocytes, lymphocytes, T-cells and B-cells, and macrophages) that go throughout the body and clean up waste and destroy germs and viruses. To do this job well, you need to supply the body with plenty of water to keep the blood and lymph from becoming thick and sluggish. Drinking adequate water, along with exercise, also helps the circulation of the lymphatic system, which bathes all the cells of the body in lymph fluid, including immune cells, such as lymphocytes and macrophages, which destroy viruses and other germs.

Drinking adequate water improves the circulation and, in turn, the work of the immune system. Water helps all of your body's cells and systems function at optimal levels. This makes good sense because over 60 percent of your body weight is water. Some organs, such as the brain, are more than 70 percent water. Whenever you become dehydrated, all of these systems suffer.

This fact is well illustrated by a study on dehydration and immunity in athletes.[65] Athletes' immune function was studied following a strenuous workout of two and a half hours. They found immunosuppression in those athletes that didn't drink enough water and who had become dehydrated, especially in the function of WBCs (neutrophils) that fight infection.

How much water do you need? Most people need at least five to eight eight-ounce glasses of water a day for good health. In hot climates or if you are exercising a lot, you need even more. The best way to see whether you are drinking adequate amounts of water is to observe the color of your urine. According to the National Kidney Foundation, you should be producing about six cups of urine daily, and the urine should be colorless or light yellow. If your urine is dark yellow, you are dehydrated and need to drink more water.[66]

To keep your immune system working at its best, follow these guidelines to ensure that you get enough water daily to stay well hydrated:

- Drink a couple of glasses of water soon after you wake up. Remember, you just went seven or eight hours with no water. The water can be warm or cold. Some people like warm or hot water early in the morning.
- Put at least two cups of water in a bottle or pitcher and make it visible and readily available in your work area. Finish it during the morning hours.
- Take liquids at lunchtime, at least the equivalent of a glass of water.
- Refill your pitcher for the afternoon and drink another two cups throughout the afternoon.
- On your way home or at supper, drink another glass of water. That makes your eight glasses.
- If you exercise or go for a walk after dinner, drink another glass or two when you get back.

By planning ahead and having water handy to drink, you will find it easier to reach your water goal daily. When I was working with a company that employed about six hundred workers, the owner was impressed with the importance of drinking water for improved performance of both the body and mind. He decided to make it easy for employees. He bought every employee a small glass pitcher with a cup for a lid. It held three cups of water. He told them to fill it in the morning and again in the afternoon and drink it all. It was a great health strategy.

Key benefits of drinking water

A recent article, "Drinking Water Quality: How to Boost Your Health and Immunity," summarized five key benefits of drinking adequate water daily:

1. *Improved energy.* When you are feeling tired, stop, take a break, and drink a glass of water.
2. *Better brain health.* Water supports circulation to the brain. One of the first signs of dehydration is mental fatigue and even a headache. Staying hydrated can help your brain function at its peak and may even help your mood. When hydrated, you simply feel better.
3. *Improved heart health.* When you do not drink adequate water, your blood thickens, making it harder for it to flow throughout your body and bring the oxygen and nutrients needed by all the cells. This increases your blood pressure and forces your heart to work harder. Staying hydrated makes it easier for your heart to do its job. One surprising finding of the Adventist Health Study from Loma Linda University was that people who drank at least five to eight glasses of water daily, in comparison to those drinking only two glasses a day, cut their risk of a fatal heart attack in half (54 percent decrease) for men and a 41 percent decrease for women.[67] That's an amazing difference, resulting from simply drinking more water.
4. *Better immune function and weight control.* Immune cells need the right nutrients to function well. Staying hydrated improves nutrient uptake and delivery to all the cells of the body. In addition, drinking water throughout the day helps curb your appetite. Instead of a snack, take a cool drink of water.
5. *Drinking water cleanses the body* from the inside out by improving kidney function and by sweating. When the body has plenty of water, the kidneys are aided in flushing away toxins. All the body's cells benefit from purified blood.[68]

Water is also important for keeping the respiratory system healthy.

This system is the avenue by which many germs or viruses enter the body. When you stay adequately hydrated, the linings of your nose, throat, and lungs also stay moist and form mucus to trap viruses and other foreign invaders from entering. When the mucus is moist, it not only traps germs better but also helps them move out of the lungs more quickly. This process is referred to as innate immunity, stopping the spread of infection. In addition to drinking water, humidifiers can also help keep the respiratory system healthy in dry climates by putting more water into the air you breathe.

TRUST IN DIVINE POWER

The last item mentioned in the list of God's natural remedies is "trust in divine power." Everyone needs hope and trust in their lives—something or someone to believe in. Even American currency, if you have not noticed lately, says, "In God We Trust."

Loma Linda University uses the healing ministry of Christ as its model for providing hope and compassion to all its patients, directing them to trust in Christ, the Divine Healer. Doctors can provide medicines, surgery, and procedures that lead to healing. But it is trust in the Divine Healer, the One who made our bodies with its healing capacity, that brings healing. Loma Linda University's motto is "To make man whole."[69] That is, to assist in the healing and restoration of the body, mind, and spirit.

Much research today shows the importance of prayer and trust in God in the healing process. Spiritual devotion can provide hope, peace, direction, meaning, and fulfillment in life. Without hope and without supportive, caring relationships, people simply do not recover as well.

Our lives depend on trust. When we cross a bridge, we trust that the engineers designed it correctly to carry the weight of the vehicles so that it will not collapse. When we board an airplane to fly to a destination, we trust that the pilot knows how to safely fly and land the plane. In the same way, we can trust our Creator. Every day the sun comes up to warm the world. Every day the plants grow and provide us with food. The laws of nature (gravity, light, time, etc.) are always there. We can count on them.

So it is in our spiritual life. God cares for His created children. "For I know the thoughts that I think toward you, says the Lord, thoughts of peace and not of evil, to give you a future and a hope" (Jeremiah 29:11). God wants us to trust Him. He has good plans for our lives and provides special help in times of difficulty.

It is easy to experience the natural laws of God. With every step we take, we experience gravity. When the light of the sun hits our back, we feel its warmth. When it comes to the desire to change our lives, the decision to make better choices in daily living, and the strength and resolve to face hardship or illness, it may be more difficult to explain God's hand at work. But what we can see and experience are the positive outcomes. It is God's power that gives us the desire to forgive, help someone even if he or she is not appreciative, and trust that everything will turn out OK, even when things look bad at the moment. Trusting in God allows us to have hope and peace, even when bad things happen in life.

Paul reminds us of this trust in the oft-quoted Romans 8:28: "We know that all things work together for good to those who love God." Paul knew by experience that out of every difficult situation, if we trust God, something good can come. Remember the time he and Silas were beaten and thrown into prison for preaching the gospel? Their legs were fastened in irons. Things looked bad for their future. But what did they do? Instead of complaining or feeling sorry for themselves, they decided to sing praises to God. They trusted Him with their lives. God brought an earthquake to set them free, and the jailer and all of his family accepted God's salvation out of this experience (Acts 16:16–34). Paul and Silas did not know what would happen when they were jailed, but they trusted God that it would all work out, and it did, far beyond their expectations.

Whenever we face challenges, whether they are physical illnesses or we need help in making lifestyle changes (losing weight, eating more healthfully, being more active, etc.), God's power and encouragement are available if we ask Him. When we are stressed and feel overwhelmed, Jesus invites us to come to Him for help. He says, "Come to me, all you who are weary and burdened, and I will give you rest. Take my yoke upon you and learn from me, for I am gentle and humble in heart, and you will find rest for your souls. For my yoke is easy and my burden is light" (Matthew 11:28–30, NIV). God will give us peace of mind and hope, even in the midst of trouble.

Benefits of trusting in God

Trusting in God brings several benefits.

Trust in God improves your outlook. Trusting in God takes away worry and gives you peace and hope for the future. It changes your attitude to have a more positive outlook on life. It is the difference between looking at a glass of water and calling it half full instead of half empty. You can choose either to focus on the problems in life or to express gratitude for the blessings you have. You have heard the story about the man who complained because his feet hurt until he met the man with no feet.

Research shows that people who have a positive outlook on life and expect good things to happen cope better with life. One group of researchers interested in the effect of optimism on health outcomes tested a group of 839 people with the Minnesota Multiphasic Personality Inventory (MMPI) to determine who were the optimists and who were the pessimists. Then they simply waited thirty years to see who was still alive and well. They found that for every ten-point increase on the pessimism scale (expecting bad things to happen in their lives), people's mortality rates increased by 19 percent. As optimism scores increased, so did their life expectancy.[70]

Another large study looked at optimism in more than seventy thousand women over a ten-year period. Compared to those with the lowest optimism scores, here are the percentages of health problems in those with the top optimism scores (top 25 percent):

- 38 percent fewer heart disease deaths,
- 39 percent fewer strokes,
- 37 percent fewer respiratory deaths,
- 52 percent fewer deaths from infections,
- 16 percent fewer deaths from cancer, and
- 29 percent fewer deaths overall.[71]

Does attitude matter? It certainly does when it comes to your health. Notice that the greatest difference had to do with the immune system. Optimists cut their risk of infections by more than 50 percent. If you want a healthy immune system, it is important to have a positive outlook on life and hope for the future that comes from trusting in God. Solomon wrote years ago about the importance of mental attitude: "Being cheerful keeps you healthy. It is slow death to be gloomy all the time" (Proverbs 17:22, GNT). Trust God and count your blessings, not your problems.

Trust in God brings strength and resolve for making lifestyle improvements. Inner strength and resolve are essential to success in making lifestyle changes. It is not easy to lose excess weight, stick with a new exercise program, or stop smoking or drinking. Trust in God increases your desire to change and provides the inner strength to make the needed changes. You can pray for what God has already promised in Philippians 4:13: "I can do all things through Christ who strengthens me." Even if achieving your goal seems impossible, trusting and depending on God can make the difference.

The Twelve-Step program of Alcoholics Anonymous (AA) recognizes that those trying to overcome their dependence on alcohol must trust in God for help. Step three says, "Made a decision to turn our will and our lives over to the care of God as we understood Him."[72] When people reach their limit, they recognize that they need God's help. They cannot do it alone. Learning to trust in God for daily help is a key teaching that has made the AA program so successful. The same principle applies when trying to lose weight, stop smoking, or committing to a regular exercise program. In Isaiah 41:10, God promises to help us: "So do not fear, for I am with you. . . . I will strengthen you and help you; I will uphold you with my righteous right hand" (NIV).

Trust in God brings comfort. When we do get sick, and things do not go well, it is good to have things that bring comfort to your life. It is easy to be discouraged when not feeling well. One comforting verse is Psalm 43:5:

Why am I so sad?
Why am I so troubled?
I will put my hope in God,
and once again I will praise him,
my savior and my God (GNT).

Prayer and trust in God bring comfort and hope, and they also help in the healing process. Even if we lose loved ones, there is comfort in knowing that God will raise all those who loved the Lord in the last days, and we can be reunited with our loved ones (1 Thessalonians 4:16, 18).

Building trust in God

We build our trust in God by getting to know Him better. Here are ways to become better acquainted with God so we can trust him more.

Talk to God. From the beginning of Creation, God met and talked with Adam and Eve. While we cannot meet and talk face-to-face with Him anymore, we can talk to God in prayer. We get to know God by getting better acquainted with His Son, Jesus Christ. John 17:3 records Jesus praying, "And this is eternal life, that they may know You, the only true God, and Jesus Christ whom You have sent."

Prayer is simply talking to God as to a friend. Let Him know your concerns, your worries, and your needs. Also, thank Him for all the blessings in your life. Jesus wants to have a relationship with you. In Revelation 3:20, Jesus invites us to take time to talk to Him: "Behold, I stand at the door and knock. If anyone hears My voice and opens the door, I will come in to him and dine with him, and he with Me." He wants to be our Friend. He wants to help us and encourage us on life's journey.

Get to know God through His Word. I like to spend time every morning reading the Word of God. God can speak to us through His Word, the Scriptures. I especially like to read the Gospels, which tell about the life of Jesus. There is no better way to learn more about God and how to trust Him than by studying the life of Christ.

Listen to God's voice speaking to your heart. God speaks to our hearts and minds through the Holy Spirit if we open our minds to Him. For example, as we read God's Word, the Spirit speaks to us, encouraging us and strengthening us. He makes God's Word come alive as we read it. Isaiah 30:21 tells us, "Your ears shall hear a word behind you, saying, 'This is the way, walk in it.' " Listen for His guidance as you read the Bible and when facing difficult decisions.

Get to know God through His created works. There is no better way to appreciate the greatness of God than to spend time in His creation. At night, as you look up at the vast universe, remember that He created it all. When you see a beautiful flower, you realize that God loves beauty. John Muir, the great naturalist, spent most of his life in nature, enjoying its great beauty. In 1912, he made the observation, "Everybody needs beauty as well as bread, places to play in and pray in, where Nature may heal and cheer and give strength to body and soul alike."[73] Nature not only testifies of God's love of beauty but also of His love for us. All of His creation was to provide a home of beauty where humankind could live in peace and harmony.

Get to know Him through fellowship with other Christians. God can speak to us through the lives of other people (e.g., by hearing their experiences and how God helped them through difficult times). One of the great blessings of attending church is not only listening to Bible teachings but visiting with other Christians. One of life's best sermons is getting to know a loving and lovable Christian—seeing love in action. Surround yourself with Christian friends. Church attendance may even help you live longer. Vanderbilt University and the University of California, Los Angeles, School of Medicine studied 5,449 people, ages forty to sixty-five, over fourteen years to see whether church attendance affected their health. They found that people who attended church once a week or more had a 55 percent lower risk of dying compared to those who did not attend.[74] Those who attend church regularly have found that trusting God benefits their lives.

Take time to help others. One of the best ways to strengthen your own faith and experience is to get involved in helping others. Philippians 2:4 says, "Let each of you look not only to his own interests, but also to the interests of others" (ESV). By giving of yourself to help others, our own lives are enriched and blessed. Jesus said it is more blessed to give than to receive (see Acts 20:35).

In Matthew 25:31–46, Jesus describes the final judgment and informs us that those He intends to take to His new kingdom are those who looked after the needs of other people. Jesus says, "For I was hungry and you gave me food, I was thirsty and you gave me drink, I was a stranger and you welcomed me, I was naked and you clothed me, I was sick and you visited me, I was in prison and you came to me." In response to the question, When did we do these things? Jesus replies, "Truly, I say to you, as you did it to one of the least of these my brothers, you did it to me" (verses 35, 36, 40, ESV). "Come, you who are blessed by my Father, inherit the kingdom prepared for you" (verse 34, ESV). As we take time to help others, we get to know God and His love firsthand.

God reprimanded the people in Isaiah's day who wanted to impress Him with their loyalty by fasting and self-denial. Then He explains true fasting:

> "Share your food with the hungry,
> and give shelter to the homeless.
> Give clothes to those who need them,
> and do not hide from relatives who need your help" (Isaiah 58:7, NLT).

God tells us that when we really show practical love to those in need, our own "health shall spring forth speedily" (verse 8, KJV). He goes on to tell us that we can call on the Lord, and He will hear us, guide us, and satisfy our souls (verses 9–11). In other words, by caring for others, our own lives and health are also blessed.

As we get to know God better and practice His way of love and caring, we also learn to trust Him. As we trust Him and His ways, we experience better health physically and spiritually.

---

1. "Mortality Risk of COVID-19: Case Fatality Rate of COVID-19 by Pre-

existing Health Conditions," Our World in Data, last updated November 18, 2020, https://ourworldindata.org/mortality-risk-covid#case-fatality-rate-of-covid-19-by-preexisting-health-conditions.

2. Tiernan Ray, "NYU Scientists: Largest US Study of COVID-19 Finds Obesity the Single Biggest 'Chronic' Factor in New York City's Hospitalizations," ZDNet, April 12, 2020, https://www.zdnet.com/article/nyu-scientists-largest-u-s-study-of-covid-19-finds-obesity-the-single-biggest-factor-in-new-york-critical-cases/.

3. Graziano Onder, Giovanni Rezza, and Silvio Brusaferro, "Case-Fatality Rate and Characteristics of Patients Dying in Relation to COVID-19 in Italy," *JAMA* 323, no. 18 (May 12, 2020): 1775, 1776, https://doi.org/10.1001/jama.2020.4683.

4. Ellen G. White, *The Ministry of Healing* (Mountain View, CA: Pacific Press®, 1905), 127.

5. Gary E. Fraser and David J. Shavlik, "Ten Years of Life: Is It a Matter of Choice?" *Archives of Internal Medicine* 161, no. 13 (July 9, 2001): 1645–1652.

6. Dan Buettner, "The Secrets of Living Longer," *National Geographic*, November 2005.

7. Lizzie Parry, "Refreshing: Fresh Air and Sunlight Can Protect Against Coronavirus, Top Scientist Claims," *U.S. Sun*, May 14, 2020, https://www.the-sun.com/news/828884/fresh-air-sunlight-protect-coronavirus-top-scientist/.

8. Parry, "Refreshing: Fresh Air and Sunlight."

9. Parry, "Refreshing: Fresh Air and Sunlight."

10. Leslie Dietz et al., "2019 Novel Coronavirus (COVID-19) Pandemic: Built Environment Considerations to Reduce Transmission," *mSystems*, April 7, 2020, https://doi.org/10.1128/mSystems.00245-20.

11. Andrew L. Goodkind et al., "Fine-Scale Damage Estimates of Particulate Matter Air Pollution Reveal Opportunities for Location-Specific Mitigation of Emissions," *Proceedings of the National Academy of Sciences* 116, no. 18 (April 30, 2019): 8775–8780, https://doi.org/10.1073/pnas.1816102116.

12. Royal College of Physicians, *Every Breath We Take: The Lifelong Impact of Air Pollution* (London: Royal College of Physicians, 2016), accessed November 18, 2020, https://www.rcplondon.ac.uk/projects/outputs/every-breath-we-take-lifelong-impact-air-pollution.

13. "What Are the Benefits of Fresh Air?" Benenden Health, July 14, 2016, https://www.benenden.co.uk/be-healthy/lifestyle/come-outside-why-fresh-air-is-essential/.

14. "Smoking and COVID-19," World Health Organization, June 30, 2020, https://www.who.int/news-room/commentaries/detail/smoking-and-covid-19.

15. Dietz et al., "Built Environment Considerations."

16. Parry, "Refreshing: Fresh Air and Sunlight."

17. Isaac Z. Pugach and Sofya Pugach, "Strong Correlation Between Prevalence of Severe Vitamin D Deficiency and Population Mortality Rate From COVID-19 in Europe," *medRxiv*, July 1, 2020, http://doi.org/10.1101/2020.06.24.20138644v1.

18. Petre Cristian Ilie, Simina Stefanescu, and Lee Smith, "The Role of Vitamin D in the Prevention of Coronavirus Disease 2019 Infection and Mortality," *Aging Clinical and Experimental Research* 32 (May 2020): 1195–1198, https://doi.org/10.1007/s40520-020-01570-8.

19. *Webster's Revised Unabridged Dictionary*, s.v. "abstemiousness," Bible Hub, accessed November 19, 2020, https://biblehub.com/topical/a/abstemiousness.htm.

20. Cambridge Dictionary, s.v. "abstemious," accessed November 19, 2020, https://dictionary.cambridge.org/us/dictionary/english/abstemious.

21. Shelby Lin Erdman, "Obesity Increases Risk of Complications From COVID-19, Damages Vaccine Efficacy, Study Finds," CNN, August 26, 2020, https://www.cnn.com/2020/08/26/health/obesity-covid-19-increased-risks/index.html.

22. Meredith Wadman, "Why COVID-19 Is More Deadly in People With Obesity—Even If They're Young," *Science*, September 8, 2020, https://www.sciencemag.org/news/2020/09/why-covid-19-more-deadly-people-obesity-even-if-theyre-young.

23. Brenda Goodman, "Why Obesity May Stack the Deck for COVID-19 Risk," WebMD, July 14, 2020, https://www.webmd.com/lung/news/20200714/why-obesity-may-stack-the-deck-for-covid-19-risk.

24. Kevin D. Hall et al., "Ultra-processed Diets Cause Excess Calorie Intake and Weight Gain: An Inpatient Randomized Controlled Trial of *Ad Libitum* Food Intake," *Cell Metabolism* 30, no. 1 (July 2, 2019): 67–77.

25. "Findings for Lifestyle, Diet, and Disease," Loma Linda University Health, accessed November 19, 2020, https://adventisthealthstudy.org/studies/AHS-2/findings-lifestyle-diet-disease.

26. "Weight Watchers," *Good Housekeeping*, February 27, 2007, https://www.goodhousekeeping.com/health/diet-nutrition/advice/a16361/weight-watchers-diet-plan/.

27. Robert Sanders, "National Panel Doubles Recommended Amount of Daily Exercise, Suggest Public Balance Food Intake With Physical Activity and Avoid Unhealthy Forms of Dietary Fat," UC Berkeley Campus News, September 5, 2002, https://www.berkeley.edu/news/media/releases/2002/09/05_diet.html.

28. World Health Organization, "Alcohol and COVID-19: What You Need to Know," April 7, 2020, https://www.euro.who.int/__data/assets/pdf_file/0010/437608/Alcohol-and-COVID-19-what-you-need-to-know.pdf; emphasis in the original.

29. "Alcohol," World Health Organization, September 21, 2018, https://www.who.int/news-room/fact-sheets/detail/alcohol.

30. National Sleep Foundation, "How Sleep Affects Your Immunity," SleepFoundation.org, accessed November 19, 2020, https://www.sleepfoundation.org/articles/how-sleep-affects-your-immunity.

31. National Sleep Foundation.

32. Maria Cohut, "How Sleep Can Boost Your Body's Immune Response," Medical News Today, February 13, 2019, https://www.medicalnewstoday.com/articles/324432.

33. Andrew Goliszek, "How Stress Affects the Immune System," *Psychology Today*, November 12, 2014, https://www.psychologytoday.com/us/blog/how-the-mind-heals-the-body/201411/how-stress-affects-the-immune-system.

34. Mayo Clinic Staff, "Sleep Tips: 6 Steps to Better Sleep," Mayo Clinic, April 17, 2020, https://www.mayoclinic.org/healthy-lifestyle/adult-health/in-depth/sleep/art-20048379.

35. Jessica Chubak et al., "Moderate-Intensity Exercise Reduces the Incidence of Colds Among Postmenopausal Women," *American Journal of Medicine* 119 (2006): 937–942, https://doi.org/10.1016/j.amjmed.2006.06.033.

36. Ellen G. White, *Healthful Living* (Battle Creek, MI: Medical Missionary Board, 1898), 130.

37. White, 131.

38. White, 132.

39. Richard J. Simpson et al., "Can Exercise Affect Immune Function to Increase Susceptibility to Infection?" abstract, *Exercise Immunology Review* 26 (2020): 8–22.

40. Giacomo Grasselli et al., "Risk Factors Associated With Mortality Among Patients With COVID-19 in Intensive Care Units in Lombardy, Italy," *JAMA Internal Medicine* 180, no. 10 (July 15, 2020): 1345–1355, https://doi.org/10.1001/jamainternmed.2020.3539.

41. F. B. Hu et al., "Walking Compared With Vigorous Physical Activity and Risk of Type 2 Diabetes in Women: A Prospective Study," *JAMA* 282, no. 15 (October 20, 1999): 1433–1439, https://doi.org/10.1001/jama.282.15.1433.

42. Anthony J. Laverty et al., "Active Travel to Work and Cardiovascular Risk Factors in the United Kingdom," *American Journal of Preventive Medicine* 45, no. 3 (September 1, 2013): 282–288, https://doi.org/10.1016/j.amepre.2013.04.012.

43. Linda S. Pescatello et al., "Exercise and Hypertension," *Medicine and Science in Sports and Exercise* 36, no. 3 (March 2004): 533–553, https://doi.org/10.1249/01.mss.0000115224.88514.3a.

44. Yechiam Ostchega et al., "Hypertension Prevalence Among Adults Aged 18 and Over: United States, 2017–2018," *NCHS [National Center for Health Statistics] Data Brief* 364 (April 2020), https://www.cdc.gov/nchs/products/databriefs/db364.htm.

45. "Health and Economic Costs of Chronic Diseases," National Center for Chronic Disease Prevention and Health Reform, Centers for Disease Control and Prevention, accessed December 17, 2020, https://www.cdc.gov/chronicdisease/about/costs/index.htm.

46. "Prevalence of Both Diagnosed and Undiagnosed Diabetes," Centers for Disease Control and Prevention, accessed December 21, 2020, https://www.cdc.gov/diabetes/data/statistics-report/diagnosed-undiagnosed-diabetes.html

47. "Heart Disease Facts," Centers for Disease Control and Prevention, accessed December 21, 2020, https://www.cdc.gov/heartdisease/facts.htm.

48. "The Nutrition Source: Nutrition and Immunity," Harvard T. H. Chan School of Public Health, accessed November 19, 2020, https://www.hsph.harvard.edu/nutritionsource/nutrition-and-immunity/.

49. "The Nutrition Source: Nutrition and Immunity."

50. "The Nutrition Source: Nutrition and Immunity."

51. "The Nutrition Source: Nutrition and Immunity."

52. "The Nutrition Source: Nutrition and Immunity."

53. Ellen G. White, *The Ministry of Healing* (Mountain View, CA: Pacific Press®, 1905), 296.

54. White, 316.

55. "The Nutrition Source: Nutrition and Immunity."

56. E. J. Mundell, "Could Zinc Help Fight COVID-19?" WebMD, September 23, 2020, https://www.webmd.com/lung/news/20200923/could-zinc-help-fight-covid-19.

57. C-H. Cheng et al., "Vitamin B_6 Supplementation Increases Immune Responses in Critically Ill Patients," *European Journal of Clinical Nutrition* 60, no. 10 (October 2006):

1207–1213, https://www.nature.com/articles/1602439.

58. Office of Dietary Supplements, "Vitamin B6: Fact Sheet for Health Professionals," National Institutes of Health, updated February 24, 2020, https://ods.od.nih.gov/factsheets/VitaminB6-HealthProfessional/.

59. Adrian R. Marineau et al., "Vitamin D Supplementation to Prevent Acute Respiratory Tract Infections: Systematic Review and Meta-analysis of Individual Participant Data," *British Medical Journal* 356 (February 15, 2017): i6583, https://doi.org/10.1136/bmj.i6583.

60. Zhila Maghbooli, Michael F. Holick et al., "Vitamin D Sufficiency, a Serum 25-Hydroxyvitamin D at Least 30 ng/mL Reduced Risk for Adverse Clinical Outcomes in Patients With COVID-19 Infection," *PLOS ONE*, September 25, 2020, https://doi.org/10.1371/journal.pone.0239799.

61. Maghbooli, Holick et al.

62. Kimberly Y. Z. Forrest and Wendy L. Stuhldreher, "Prevalence and Correlates of Vitamin D Deficiency in US Adults," *Nutrition Research* 31, no. 1 (January 2011): 48–54, https://doi.org/10.1016/j.nutres.2010.12.001.

63. Ga Young Lee and Sung Nim Han, "The Role of Vitamin E in Immunity," *Nutrients* 10, no. 11 (November 2018): 1614, https://doi.org/10.3390/nu10111614.

64. Lee and Han.

65. Takeharu Chishaki et al., "Effects of Dehydration on Immune Functions After a Judo Practice Session," *Luminescence: The Journal of Biological and Chemical Luminescence* 28, no. 2 (2013): 114–120, https://doi.org/10.1002/bio.2349.

66. "6 Tips to Be 'Water Wise' for Healthy Kidneys," National Kidney Foundation, accessed November 23, 2020, https://www.kidney.org/content/6-tips-be-water-wise-healthy-kidneys.

67. Jacqueline Chan et al., "Water, Other Fluids, and Fatal Coronary Heart Disease: The Adventist Health Study," *American Journal of Epidemiology* 155, no. 9 (May 1, 2002): 827–833, https://doi.org/10.1093/aje/155.9.827.

68. "Drinking Water Quality: How to Boost Your Health and Immunity," Water Right Group, September 4, 2020, https://www.water-rightgroup.com/blog/water-quality-boost-health-and-immunity/.

69. "University Philosophy," Loma Linda University, accessed November 23, 2020, http://llucatalog.llu.edu/about-university/university-philosophy/.

70. Toshihiko Maruta et al., "Optimists Vs Pessimists: Survival Rate Among Medical Patients Over a 30-Year Period," *Mayo Clinic Proceedings* 75, no. 2 (February 1, 2000): 140–143, https://doi.org/10.4065/75.2.140.

71. Eric S. Kim et al., "Optimism and Cause-Specific Mortality: A Prospective Cohort Study," *American Journal of Epidemiology* 185, no. 1 (January 2017): 21–29, https://doi.org/10.1093/aje/kww182.

72. Alcoholics Anonymous, *Alcoholics Anonymous*, 4th ed. (New York: Alcoholics Anonymous Worldwide Services, 2001), 59.

73. John Muir, *The Yosemite* (New York: Century, 1920), 256.

74. Marino A. Bruce et al., "Church Attendance, Allostatic Load and Mortality in Middle Aged Adults," *PLOS One*, May 16, 2017, https://doi.org/10.1371/journal.pone.0177618.

Step 4

Get Good Medical Care

We've talked about the importance of avoiding contact with germs and viruses that cause infections and the importance of maintaining a healthy immune system to prevent infectious disease and to survive the virus should you become infected. We've talked about living a healthy lifestyle to prevent chronic diseases. In addition to these strategies, it is also important to get good medical care. Your doctor can play a vital role in helping you survive infections and serious diseases. For example, in a study of 1.3 million confirmed coronavirus cases, those who had a chronic health condition were at significantly higher risk for severe complications or mortality than those without underlying health problems.[1] According to the study, the three most common underlying health conditions were cardiovascular disease (32 percent of infected patients), diabetes (30 percent of patients), and chronic lung disease (18 percent of patients). Hospitalization for those with underlying conditions was six times higher, and mortality was twelve times higher than those without underlying health problems. Not having any chronic health condition certainly improves your survival chances.

The most common chronic diseases in the United States are heart disease, cancer, chronic lung disease, stroke, Alzheimer's disease, diabetes, and chronic kidney disease. Six in ten of all adults in the US have one of these chronic diseases; four in ten have two or more.[2] If you have one of these problems, it is imperative that you work with your doctor to get these conditions under control and properly managed. Follow your doctor's guidance: take medications if prescribed, get regular checkups and preventive exams, and make needed lifestyle changes. The better that chronic diseases are managed, the better your chances for survival and longevity. The good news is that most of these conditions can be prevented by implementing a healthy lifestyle. Here are concise steps you can take to help manage common health problems you may have.

Diabetes or high blood sugar

- Lose excess weight. Even losing ten or twenty pounds can be very helpful.
- Exercise daily, such as brisk walking, for a minimum of thirty minutes or as your doctor recommends.
- Avoid soft drinks, sweets, refined foods (chips, pastries, white bread, white rice, etc.), and foods high in saturated fat (red meat, ice cream, butter, cheese, etc.).
- Eat primarily whole plant-based foods (vegetables, salads, beans, lentils, soy foods, fresh fruits, whole grains, nuts, and other plant proteins) in moderate amounts to achieve a healthy weight.
- Don't smoke.
- Take medications if prescribed.
- Keep A1C blood tests (a measure of blood sugar) less than 7.0 or as your doctor prescribes.
- Keep stress levels moderate and get adequate rest and sleep daily.

Coronary heart disease or high cholesterol

- Lose excess weight.
- Exercise regularly, thirty minutes or more daily, such as brisk walking, biking, or swimming, or as your doctor recommends.
- Eat healthy meals, primarily whole plant foods, and avoid or limit meats and other foods that are high in saturated fat. To help lower cholesterol, eat high-fiber foods, such as whole grains, including oatmeal; peas; beans; lentils; garbanzos; soy; vegetables; fruits; and nuts. They are all good for the heart.
- Don't smoke.
- Keep blood pressure in a healthy range (ideally less than 120/80) or as your doctor directs.
- Keep blood cholesterol levels in a healthy range (LDL cholesterol less than 100 mg/dL). If prescribed, take a cholesterol-lowering medicine to achieve your goal, along with the actions listed above.

Cancer

- If you have cancer, follow your doctor's directions and medications as prescribed.
- Include healthy eating (more whole plant-based foods, less red meat and refined foods) in moderation to avoid obesity.
- Be as active as you are able; follow your doctor's guidance.
- Get adequate rest and sleep daily.
- Trust in divine power to help you heal and recover.

High blood pressure

- Maintain a healthy weight. Losing even ten or fifteen pounds helps considerably.
- Limit salt (sodium) in the diet as much as possible (less than 2,400 milligrams of sodium daily).
- Eat more fruits and vegetables. The DASH diet by the National Institutes of Health recommends as many as seven to nine servings daily (they are low in calories, high in fiber, low in sodium, and high in potassium—all of which help lower blood pressure).
- Get moderate exercise daily, thirty-plus minutes of brisk walking, biking, or swimming, or as your doctor directs.
- Aim for a blood pressure less than 120/80 for best health, or as your doctor prescribes.
- Avoid alcohol.
- Don't smoke.
- Eat healthy meals, low in saturated fat and refined foods. Choose primarily whole plant-based foods high in fiber.
- If needed, take medication to lower blood pressure.

Lung disease

- Don't smoke.
- If you have asthma, bring it under control with medication if needed.
- Be as active as you can, following your doctor's guidance.
- Eat healthy meals.

• Follow your doctor's guidance for managing your lung disease.

As you can see, God's natural remedies are the foundation of a healthy life and are important for helping control or prevent nearly all major health problems today, including lowering your risk for infectious disease.

Medical treatment

Should you develop any health problem, you must work closely with your doctor because the doctor can do much to treat the condition, limit serious complications, and avoid premature death. Treatments for all diseases continue to get better. More people are surviving. Getting early treatment also improves outcomes. We can be thankful that we live in a country where we have good treatment options.

Keep in mind that most cases of coronavirus were not reported, and the actual numbers may be as high as ten times the confirmed cases, so the total mortality rate was actually much lower.[3] The CDC estimates the actual 2020 case-fatality rate in the US was about 0.65 percent. This varied by age group, of course. The infection mortality rate (IFR) for people less than twenty years old was only 0.003 percent, and the IFR for people twenty to forty-nine was 0.02 percent (only two people out of ten thousand). The IFR was 0.5 percent for fifty- to sixty-nine-year-old persons. The IFR only rose above one death per hundred people infected (5.4 percent) in those ages seventy and older. It is encouraging to see that mortality rates for the majority of the population—those under seventy—were quite low. This also means, however, that those seventy and older need to be the most careful.

It is important to note that for every two deaths from COVID-19, there was an additional death caused by the consequences of the pandemic. There was an escalation in deaths caused by suicide, drug overdose, and not getting adequate care for chronic diseases. This means an additional hundred thousand lives or more were lost.[4] Alcohol intake also increased significantly during the pandemic lockdown. According to a National Institute of Alcohol Abuse and Alcoholism survey, people drank in larger quantities and on more days of the week. The number of women binge drinking increased by 41 percent.[5] Furthermore, the

CDC lists alcohol intake among the top four chronic diseases in the US.[6]

In 2020, COVID-19 captured our attention due to its sudden and widespread presence. However, it is helpful to compare all causes of death in the US.

Cause of death	Deaths/day in the US	Percentage of all deaths
All causes of death	7,970 (2019)	
Heart disease	1,773	22.2
Cancer	1,641	20.6
COVID-19 (as of November 9, 2020)	580★	7.2
Accidents	465	5.8
Lung disease (excluding COVID-19)	438	5.5
Stroke	401	5.0
Alzheimer's disease	332	4.2
Flu & pneumonia (2019–2020 season)	331	1.9
Diabetes	228	2.9
Drug overdose (higher in 2020)	192 (2019)	2.4
Suicides (higher in 2020)	132	1.7
Homicides	53	0.7
Lifestyle risks		
Poor diet	1,835	---
Smoking	1,205	---
Obesity	1,027	---
Sedentary lifestyle	658	---
Alcohol use	260	---

★ Average deaths per day based on a full year. Actual deaths per day from February to October 26, 2020, was 822. This figure varies from day to day and will continue to change as more people are vaccinated.

Sources: "Leading Causes of Death," National Center for Health Statistics, CDC; US Burden of Disease Collaborators, "The State of US Health, 1990–2010," *JAMA* 310, no. 6 (2013): 591-606; CDC COVID Data Tracker.

When all the news was focusing on coronavirus deaths, it seemed to overwhelm us. It's important to understand that many other health problems result in just as many or even more deaths. More people die each day from a poor diet, smoking, obesity, and a sedentary lifestyle.

That's why it's important to focus on God's natural remedies that help lower the risk for *all* major health problems, including infectious disease. It's also good to know that new treatments for disease are always coming available. Also, case-fatality rates decrease remarkably as immunizations increase among the population. Take courage!

IMMUNIZATION

Vaccines are one of the most effective strategies to protect against serious infections. When a vaccine is given, it stimulates the immune system—both T-lymphocytes, which recognize an invading virus, and B-lymphocytes, which produce antibodies to fight the virus. Thus, vaccines boost the body's natural defenses to safely develop immunity to a specific disease.

Also, after you recover from an infection, this immune response activates to help prevent a reoccurrence of the disease. This process takes several days, however. Once immunity has developed, when you encounter viruses from exposure, your immune system quickly reacts to attack the invaders and prevents the infection from taking over.

Vaccines develop the body's immunity by imitating an infection. The vaccine, however, seldom causes illness. At times, after getting a vaccine, some people's immune response may cause minor symptoms, such as lethargy or maybe even a fever. Such minor symptoms are normal and should be expected as the body builds immunity. Once the "imitation infection" goes away, the body has developed immunity that can protect you against the dangerous potential of the real virus infection. Most people have few or no symptoms.

Some vaccines require two doses to increase immunity most effectively. If needed, the second dose—commonly called a booster shot—helps bring immunity up to the level needed to be most effective.

Some viruses mutate, and the immune system doesn't recognize them anymore. For these viruses, you need a vaccine every year for it to recognize the mutated virus. Flu shots fall in this category.[7]

Questions have been raised about the safety of some vaccines. An article in the journal *American Family Physician* addressed this issue. It states that fever and irritability can occur after some vaccines, as well as redness and soreness at the site of the injection. In rare cases, more serious complications might occur. The article also states, "Currently, no substantial evidence links measles-mumps-rubella vaccine to autism, or hepatitis B vaccine to multiple sclerosis. Thimerosal [with mercury] is being eliminated from routine childhood vaccines because of concerns that multiple immunizations with vaccines containing this preservative might exceed recommended mercury exposures."[8] You can ask for a flu vaccine with no thimerosal added. It is available for those who are concerned.

Vaccinations have been one of the most successful strategies against infectious disease, literally saving thousands of lives. Smallpox, once a deadly disease, killed an estimated three hundred million people in the twentieth century alone. This disease killed one out of every three people it infected. The disease was conquered only by vaccinations.

Polio was another major killer in my lifetime. Some readers may remember when they were children that they were afraid to go to public meetings or go swimming in the city swimming pool because people thought they might get polio. My mother got polio as a young girl and nearly died. It left her with a serious problem walking. She did quite well in her midlife years, but as she got older, she had to spend the last eight years of her life in a wheelchair. I remember when I got my polio vaccine and was so glad I was protected against that dreaded disease. I remember pictures of polio victims in iron lungs to breathe for them. As a young child, I found it very scary. Thankfully, America was declared free of wild poliovirus in 1994 due to the effectiveness of vaccinations.

Here are some other common infectious diseases that have been nearly eliminated with vaccines.

Measles	894,134 cases in 1941	86 cases in 2000
Mumps	152,209 cases in 1968	338 cases in 2000
Rubella	12 million cases 1964–1965	176 cases in 2000

Some of these common diseases could be fatal or have serious complications. One to two people out of a thousand with measles died; a similar number got encephalitis. Life could not be conducted as normal without safe, effective vaccines.

The CDC and the FDA (Food and Drug Administration) oversee the safety and effectiveness of new vaccines. Vaccines are not released for public use until proven safe and have at least a 50 percent effectiveness rate for preventing a disease. Most vaccines run from 70 to 95 percent effective. (There are no guarantees in life, other than taxes and death.) Even when we use a car for transportation, we are at increased risk. Some 38,800 people died in automobile accidents in 2019.[9] Yet most of us still drive daily. Any risk from a vaccine is extremely small compared to the potential of preventing hundreds of thousands of deaths every year. We have to evaluate the benefit to risk and decide what provides the best option for a healthy, normal life. If you have concerns about a new vaccine's safety, talk to your doctor.

My son is a physician, and during the COVID-19 pandemic, he eagerly awaited vaccines to protect all of his office workers. Because they were required to deal with infected patients daily, they were at high risk of getting the virus. Within eight months, over two hundred thousand people died from the disease in the US. Public health officials were eager to get a safe vaccine as soon as possible to protect our nation from further serious disease and more deaths.

CONCLUSION

I have outlined a four-step strategy to lower your risk of chronic and infectious disease. Let me summarize these four key steps.

1. Avoid exposure to viruses and other germs (washing hands, wearing masks, cleaning surfaces, avoiding sick people, staying in place when sick).
2. Maintain a healthy immune system by getting your rest. Make sure you eat well and have adequate levels of key nutrients. Get regular exercise. And avoid becoming fatigued or run down from doing too much, being overly stressed, or not getting adequate rest.
3. Develop a healthy lifestyle to keep your immune system functioning at peak performance and prevent or correct any existing health problems. Remember God's natural remedies. "Pure air, sunlight, abstemiousness, rest, exercise, proper diet, the use of water, trust in divine power—these are the true remedies. Every person should have a knowledge of nature's remedial agencies and how to apply them."[10]
4. Get good, regular medical care to prevent or manage any existing health problems you may have. And when available, get vaccinated to boost your immune response so that it will protect you from any preventable infectious disease.

Having taken these practical steps, you can trust the Lord that your life is safe and secure in Him, no matter what happens. In conclusion, my wish for you is well expressed in 3 John 1:2: "Beloved, I pray that you may prosper in all things and be in health, just as your soul prospers."

---

1. Erin K. Stokes et al., "Coronavirus Disease 2019 Case Surveillance—United States, January 22–May 20, 2020," *Morbidity and Mortality Weekly Report* 69, no. 24 (June 19, 2020): 759–765, http://dx.doi.org/10.15585/mmwr.mm6924e2; https://www.cdc.gov/mmwr/volumes/69/wr/mm6924e2.htm?s_cid=mm6924e2_w.

2. "Chronic Diseases in America," National Center for Chronic Disease Prevention and Health Promotion, Centers for Disease Control and Prevention, accessed December 21, 2020, https://www.cdc.gov/chronicdisease/resources/infographic/chronic-diseases.htm.

3. Matt Perez, "Ten Times More People Have COVID-19 Antibodies Than Are Diagnosed, CDC Reports," *Forbes*, June 25, 2020, https://www.forbes.com/sites/mattperez

/2020/06/25/ten-times-more-people-have-covid-19-antibodies-than-are-diagnosed-cdc-reports/.

4. Misha Gajewski, "What's the True Toll of the Coronavirus Outbreak? COVID-19 Deaths Aren't the Only Reason for the Rising Death Rates," *Forbes*, October 12, 2020, https://www.forbes.com/sites/mishagajewski/2020/10/12/whats-the-true-toll-of-the-coronavirus-outbreak-covid-19-deaths-arent-the-only-reason-for-the-rising-death-rates/.

5. Sasha Pezenik, "Alcohol Consumption Rising Sharply During Pandemic, Especially Among Women," ABC News, September 29, 2020, https://abcnews.go.com/US/alcohol-consumption-rising-sharply-pandemic-women/story?id=73302479.

6. "Chronic Diseases in America," Centers for Disease Control and Prevention, last updated September 24, 2020, https://www.cdc.gov/chronicdisease/resources/infographic/chronic-diseases.htm.

7. As of the date of this writing, not enough was known about the COVID-19 virus to know whether a vaccine for it would need to be administered just once or yearly.

8. Sanford R. Kimmel, "Vaccine Adverse Events: Separating Myth From Reality," *American Family Physician* 66, no. 11 (December 1, 2002): 2113–2121, https://www.aafp.org/afp/2002/1201/p2113.html.

9. "Motor Vehicle Deaths Estimated to Have Dropped 2% in 2019," National Safety Council, accessed November 23, 2020, https://www.nsc.org/road-safety/safety-topics/fatality-estimates.

10. Ellen G. White, *The Ministry of Healing* (Mountain View, CA: Pacific Press®, 1905), 127.

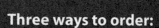